iRODS
User Group Meeting 2015
Proceedings

7TH ANNUAL CONFERENCE SUMMARY

The iRODS User Group Meeting of 2015 gathered together iRODS users, Consortium members, and staff to discuss iRODS-enabled applications and discoveries, technologies developed around iRODS, and future development and sustainability of iRODS and the iRODS Consortium.

The two-day event was held on June 10th and 11th in Chapel Hill, North Carolina, hosted by the iRODS Consortium, with over 90 people attending. Attendees and presenters represented over 30 academic, government, and commercial institutions.

Contents

Listing of Presentations

The following presentations were delivered at the meeting:

- **The State of the iRODS Consortium**
  Dan Bedard, University of North Carolina-Chapel Hill

- **iRODS Development Update, 4.0.2 to 4.1.0**
  Jason Coposky, University of North Carolina-Chapel Hill

- **EMC MetaLnx: iRODS Admin and metadata management interface**
  Stephen Worth, EMC Corporation

- **Implementing a Genomic Data Management System using iRODS at Bayer HealthCare**
  Carsten Jahn, Bayer Business Services, GmbH
  Navya Dabbiru, Tata Consulting Services

- **An Open and Collaborative Enterprise Software Platform for Patient and Compound Profiling**
  Marc Flesch, Genedata AG

- **Use Cases with iRODS and CiGri at CIMENT**
  Bruno Bzeznik, Université Joseph Fourier-Grenoble

- **iRODS in Complying with Public Research Policy**
  Vic Cornell, DataDirect Networks

- **iRODS at the Sanger Institute**
  Jon Nicholson, Wellcome Trust Sanger Institute

- **Secure Cyberinfrastructure for Open Science**
  Chris Jordan, University of Texas at Austin

- **The iPlant Data Commons: Using iRODS to Facilitate Data Dissemination, Discovery, and Reproduction**
  Tony Edgin, iPlant Collaborative/University of Arizona

- **Prospects and Needs for iRODS at CC-IN2P3**
  Jean-Yves Nief, CC-IN2P3

- **YODA: Easy access and management of research data across universities**
  Ton Smeele, Utrecht University

- **Best practices in construction of a nationwide iRODS grid**
  Tom Langborg, Linköping University

- **iRODS Roadmap, 4.2.0 and Beyond**
  Jason Coposky, University of North Carolina-Chapel Hill

- **First Look: iRODS Web Interface Development**
  Mike Conway, University of North Carolina-Chapel Hill

- **Java Development for iRODS**
  Mike Conway, University of North Carolina-Chapel Hill

- **Data Management challenges in Today's Healthcare and Life Sciences Ecosystem**
  Jose L. Alvarez, Seagate Cloud and Systems Solutions

- **HydroShare: Advancing Hydrology through Collaborative Data and Model Sharing**
  David Tarboton, Utah State University

- **RADII: Resource Aware Data-centric collaborative Infrastructure**
  Tahsin Kabir, University of North Carolina-Chapel Hill

- **iRODS Feature Requests and Discussion**
  Reagan Moore, University of North Carolina-Chapel Hill

A panel session was held:

- **Working with iRODS at Scale**
  Tom Langborg, Linköping University
  Nirav Merchant, iPlant Collaborative/University of Arizona
  Jean-Yves Nief, CC-IN2P3

Posters presented include:

- **State of the iRODS Consortium, Summer 2015**
  Dan Bedard, University of North Carolina-Chapel Hill

- **Policy Sets – integrated Rule Oriented Data System**
  Reagan Moore, University of North Carolina-Chapel Hill

- **Practical Policy Working Group, Research Data Alliance**
  Reagan Moore, University of North Carolina-Chapel Hill

- **Reproducible Data Driven Research, DataNet Federation Consortium**
  Reagan Moore, University of North Carolina-Chapel Hill

- **Implementing a Genomic Data Management System using iRODS at Bayer HealthCare**
  Navya Dabbiru, Tata Consulting Services
  Volker Hadamschek, Bayer Business Services, GmbH
  Ulf Hengstmann, Bayer Business Services, GmbH
  Carsten Jahn, Bayer Business Services, GmbH
  Arun Kumar, Bayer Business Services, GmbH
  Thomas Leyer, Bayer Business Services, GmbH
  Uma Rao, Tata Consulting Services
  Thomas Schilling, Bayer CropScience, AG
  Henrik Seidel, Bayer Pharma, AG

- **MetaLnx – An iRODS Administrative & Metadata Management Tool**
  Stephen Worth, EMC Corporation

Articles

iRODS Cloud Infrastructure and Testing Framework

Terrell Russell
Renaissance Computing Institute
(RENCI)
UNC Chapel Hill
unc@terrellrussell.com

Ben Keller
Renaissance Computing Institute
(RENCI)
UNC Chapel Hill
kellerb@renci.org

ABSTRACT

In this paper, we describe the iRODS Cloud Infrastructure and Testing Framework, its design goals, current features, history of development, and some future work.

Keywords

iRODS, testing, automation, continuous integration

INTRODUCTION

The iRODS Cloud Infrastructure and Testing Framework aims to serve a variety of constituents. The target audiences include core developers, external developers, potential users, current users, users' managers, and grid administrators.

Testing provides confidence to the developers and the community that the codebase is stable and reliable. This assures users and users' managers that infrastructure built with iRODS can be trusted to enforce their organizations' data management policies.

The Framework currently consists of three major components: Job queue management and automation (provided by Jenkins[1]), business logic and deployment management (provided by Python and Ansible[2] modules), and VM resource management (provided by VMware's vSphere[3]). The connective code is written in Python and included in three git repositories.

These components work together to generate reproducible builds of any commit and to test those builds on a growing matrix of operating systems, their versions, iRODS versions, and a large number of combinations of the different types of plugins iRODS supports: databases, resource types, authentication types, and types of network connection.

MOTIVATION AND DESIGN GOALS

Software testing, in general, is a tool used to generate one thing: confidence. Everything else is secondary.

The way iRODS generates confidence is through the following design goals:

- Transparency (in both process and product)

- Use of existing industry best practices

---

[1] https://jenkins-ci.org/
[2] http://www.ansible.com/home
[3] http://www.vmware.com/products/vsphere

iRODS UGM 2015 June 10-11, 2015, Chapel Hill, NC

- Coverage → Confidence in Refactoring

- Packaging → Ease of installation and upgrade

- Test framework idempotency

- Test independence

- Topology awareness

- Automation, Automation, Automation

Using open source best practices affords a great deal of transparency. The package managers built into iRODS users' operating systems allow iRODS to manage dependencies very cleanly and to reduce the amount of code in the iRODS core codebase. Having the ability to run a single test with the knowledge that it will leave the system in a clean state allows for a tremendous reduction in the amount of time it takes to run through the "code-deploy-test" development cycle. Ideally, the tests are smart enough to only run when the iRODS Zone being tested matches the right criteria (standalone server or a full topology, with or without SSL, using the correct authentication scheme, etc.).

And finally, all of these goals only matter if everything the tests do can be automated. Without automation, testing remains more than a full-time job. With automation, it can happen alongside code development and drive other best practices.

zone_bundle.json schema

iRODS 4.1 has incorporated schema-based configuration and validation[4]. The new *izonereport* iCommand, produces a JSON representation of the entire topology of the local Zone, as well as configuration information for all the included servers, databases, resources, and available plugins. Saved as a "zone_bundle.json" file, this representation serves multiple purposes.

First, this representation affords the development team a concrete view of a particular deployment. For debugging purposes, this is paramount and significantly reduces the overhead, as well as the back and forth, of setting the context for a support request when someone is having trouble. More generally, it gives the person having trouble a chance to see the same overview before submitting it for help, which may be enough to assist them in identifying the error themselves.

Second, this representation is the same format that the Framework uses to deploy a topology into the private cloud. The VMs that are dynamically launched are configured to match those that are described in the Zone bundle. This allows for much quicker turnaround between description of a problem and real, tangible Zones that can be manipulated and tested directly.

A third potential benefit is the de facto shared interchange format for graphical representations of iRODS Zones. As graphical tools for iRODS become more sophisticated, we envision both administration and support applications getting heavy use out of this newly defined format.

Lastly, the move to schema-driven configuration management, in general, is a strong indicator of a mature product. This representation of the full configuration of a Zone should unlock powerful automation capabilities.

PROGRESSION OVER TIME

The Framework has come a long way in the last few years. While the goals have been consistent, the implementation and the technologies used have been under constant development.

---

[4]https://github.com/irods/irods_schema_configuration

July 2011

$$Python \rightarrow Node.js \rightarrow RabbitMQ \rightarrow Celery \rightarrow Eucalyptus$$

This initial effort was manually driven by a Python script that validated a configuration file and then used RabbitMQ and a Celery message bus to deploy, test on, and then destroy Eucalyptus VMs. This was a lot of machinery to keep up and running, and all the machines involved had to be running very consistent versions of Python, Erlang (for RabbitMQ), and Celery. As the entire point of the Framework was to deploy both new and old operating systems, this consistency proved overwhelming.

October 2012

$$Python \rightarrow Node.js \rightarrow ssh \rightarrow OpenStack$$

The second iteration of the Framework reduced the number of moving parts and moved to OpenStack (Diablo release) for VM provisioning. This was still a manually driven process but allowed us to add two more operating systems to our testing matrix pretty painlessly.

January 2013

$$Hudson \rightarrow Python \rightarrow OpenStack$$

The third version of the Framework served two major functions. It incorporated the RENCI-wide Hudson job manager for automation, and it removed the dependency on Node.js. This version served as the new baseline for what iRODS testing would look like for the next couple of years.

October 2013

$$Hudson \rightarrow Python \rightarrow vSphere\ long\text{-}running\ VMs$$

The fourth version was motivated by the ongoing problems we were having with the ability to add new operating system images into the available pool managed by OpenStack. The iRODS development team was not managing the OpenStack infrastructure itself, and so did not have full control over the environment. We opted to move our work from the research side to the production side of the RENCI infrastructure and, in doing so, move to a group of long-running VMs that would always be available.

CURRENT INFRASTRUCTURE

Earlier this year, the fifth version of the iRODS Cloud Infrastructure and Testing Framework came online.

Spring 2015

$$Jenkins \rightarrow Python \rightarrow Ansible \rightarrow vSphere\ dynamic\ VMs$$

This version improves both the reproducibility and robustness of the last two years of work. Moving to our own installation of Jenkins means that all the servers are controlled by the iRODS development team. The biggest difference in this version is the use of dynamically deployed (and then destroyed) vSphere VMs, which are configured via Ansible modules. Using dynamic VMs ensures that every test run starts from the same initial conditions and removes the need for "clean up" code that was required in the previous Framework iteration to return the long-running VMs to a pre-test state.

Jenkins

Jenkins is an open source continuous integration server that allows the iRODS development team to schedule and coordinate the different build and test jobs that power the Framework.

Currently, Jenkins is scheduled to build and test the master branch every thirty minutes, if there have been new commits. These tests include coverage of all the major features of iRODS that can be exercised on a single standalone server.

Separately, Jenkins has a pipeline which runs multiple flavors of topology tests (tests that require an iRODS Zone comprising multiple VMs) on multiple flavors of iRODS Zones. For each of these tests, several dynamic VMs are created, networked together, and configured to run iRODS, all based on the contents of a Zone bundle.

Automatic testing of federated Zones is next on the list of things to add. Currently, all Federation testing on the core iRODS code has been done manually. This takes a long time to configure properly and is hard to reproduce.

Python and Ansible

Ansible is an automation and management framework for executing code on multiple remote servers. It is written in Python and allows users to execute bits of Python code (Ansible modules) on remote machines and gather the results.

vSphere

VMware's vSphere is a server virtualization platform that allows the iRODS development team to programmatically stand up, interact with, and then tear down virtual servers (or virtual machines, or VMs). The RENCI infrastructure is currently configured with enough compute, memory, and storage to provide the iRODS development team the capability to have up to 100 VMs active and under management.

As the combination of iRODS configurations under testing continues to grow, the number of concurrent servers required will grow as well.

Workflow

Jenkins[5] launches a job named *build-all*. This job builds the iRODS packages on all of the currently supported operating systems and architectures. The main work of this job is handled by a *build.py*[6] script:

```
python build.py \
    --build_name ${BUILD_TAG} \
    --output_root_directory ${PACKAGES_ROOT_DIRECTORY} \
    --git_repository ${PARAMETER_GIT_REPOSITORY} \
    --git_commitish ${PARAMETER_GIT_COMMITISH}
```

When this job is complete, control is returned to Jenkins, which then fans out and launches many jobs, one for each combination of standalone test that is currently supported:

---

[5]https://jenkins.irods.org/
[6]https://github.com/irods/irods_testing_zone_bundle

standalone-test-centos6-mysql

standalone-test-centos6-oracle

standalone-test-centos6-psql

standalone-test-opensuse13-psql

standalone-test-ubuntu12-mysql

standalone-test-ubuntu12-psql

standalone-test-ubuntu14-mysql

standalone-test-ubuntu14-psql

test-jargon

The work of each of the standalone jobs is handled by a *jenkins_helper.py*[7] script which calls the following Python functions, in order: *deploy*, *enable_ssl*, *test*, *gather*, and *destroy*.

When these steps are complete, Jenkins has a full report of how each test progressed, the job's return status, as well as a set of gathered result files which are used to display the results graphically and aid in debugging failed tests. There are currently 1,253 Python tests run on each combination of operating system and database.

The work of the *test-jargon* job is handled by a separate *test.py*[8] script which runs the *irods_test_jargon* Ansible module. The Jargon test results (currently 1,917 tests) are gathered and displayed in Jenkins.

Each vSphere VM that is started and stopped is handled by the Python provisioner written to interface with vSphere[9]. The Framework creates and destroys VMs using an injected Python module (which uses pyVmomi[10]). This separates the choice of VM provisioning technology from the use of the Framework.

Coverage

The current total coverage for the core iRODS codebase is 61.5%. This number represents the C++ line coverage in the main iRODS repository, including the iRODS server(s), iCommands, and plugins.

	Additional	Cumulative	
Standalone single server	59.2%	59.2%	observed
Topology	2.3%	**61.5%**	observed
Error checking	~14%	75%	estimated
Untested features	~25%	100%	estimated

The standalone coverage results are generated by running the entire test suite on a single iRODS iCAT server with no additional Resource servers in the Zone. The standalone coverage is currently 59.2%.

The topology coverage results are generated by running the entire test suite on a Zone with four servers (1 iCAT and 3 Resource servers) two times (once each on the iCAT and then a Resource server) and then gathering and combining the coverage logs from all the servers. The total topology coverage is currently 61.5% which means the topological testing adds 2.3% of marginal code coverage.

These results include no unit tests at this time. All of the tests we run are functional tests from outside the system. We estimate, by inspection, that less than half of the remaining lines are unexercised error cases due to our lack of

---

[7] https://github.com/irods/irods_testing_zone_bundle

[8] https://github.com/irods/irods_testing_jargon

[9] https://github.com/irods/irods_testing_provisioner_vsphere

[10] https://github.com/vmware/pyvmomi

simulating database connection errors, passing malformed input to individual functions, etc. The remainder of the lines are untested features that we estimate to be about 25% of the codebase.

Increasing coverage is always a goal, but it is not something that we have explicitly targeted to date. We expect this to change in the next year.

FUTURE WORK

Looking ahead, we have a list of things we intend to add to the existing Framework.

Fall 2015

$$Jenkins \rightarrow Python \rightarrow Ansible \rightarrow zone\_bundle \rightarrow vSphere\ dynamic\ VMs$$

- Make tests "zone_bundle aware" - This will allow the tests to know everything about the Zone in which they are running and to intelligently skip certain tests. This will require pushing the zone_bundle down through Ansible.

- Move to CMake - The current build system is a cobbled combination of bash, perl, python, and make. The move to CMake will be more standardized and reduce the complexity for us as well as external developers.

- Separately versioned *external/* - The *irods/external/* directory is currently tied to the iRODS version. Separating this will allow *external/* to move at its own speed, and give the Framework full capability to compile past versions of iRODS without building *external/* from source.

- Testing unpushed branches - Expanding the Framework's functionality to include building and testing a developer's local unpushed branch will reduce the number of complex "blind" commits that are too time-consuming to test by hand.

- Enforced code review - We are planning to introduce code review (perhaps via Gerrit) to an internal server before pushing to a public server to ensure a high-quality level of commit.

- Federation testing - This will increase visibility into and confidence around updating the iRODS wireline protocol (planned for iRODS 5.0).

CONCLUSION

The iRODS Cloud Infrastructure and Testing Framework has been under development for over three years and has been through five major versions.

As a system that tests a distributed system, the Framework may be useful to others. We are encouraged by the rapid pace we have been able to recently add new features.

The current version builds and tests iRODS across a growing matrix of operating systems, their versions, iRODS versions, and a large number of combinations of the different types of plugins iRODS supports: databases, resource types, authentication types, and network types. The iRODS code is tested across both single machine deployments as well as multi-machine topologies via *zone_bundle* JSON files.

Through transparency and automation, the iRODS Cloud Infrastructure and Testing Framework provides confidence in the claims that iRODS makes as a production-ready software technology.

Data Intensive processing with iRODS and the middleware CiGri for the Whisper project

Briand Xavier*   Bzeznik Bruno†

Abstract

Like many projects in science of the universe, the seismological project Whisper is faced with massive data processing. This leads to specific IT softwares for the project as well as suitable IT infrastructures. We present here both aspects.

We provide a flexible way to design a sequence of processing. We also deal with data-management and computational optimization questions. On IT infrastructure, we present the platform CIMENT provided by the University of Grenoble. It offers a data-grid environment with the distributed storage iRODS and the grid manager CiGri.

This is the partnership between these two IT components that has enabled a data-intensive processing and, also, permitted the Whisper project to bring new scientific results.

Keywords: Data-Intensive, Grid computing, Distributed storage, Seismic Noise, Whisper, CiGri, iRODS.

1   Introduction

The *Whisper* * project is a an european project on seismology whose goal is to study properties of the earth with the seismic ambient noise such that evolution of seismic waves speed. This noise is almost all the signal continuously recorded by the seismic stations worldwide (Europe, China, USA, Japan), except earthquakes. It offers new observables for the seismologists, new types of virtuals seismograms that are not only located at the place of earthquakes and that are provided by the operation of correlation which requires significant computations. For instance, one can obtain wave paths that probes the deepest part of the Earth [2, 12].

Accordingly, this is one of the first project in the seismological community that studies systematically the continuous recordings, which represents a large amount of seismological data, of the order of several tens of terabytes. For instance, one year of the Japanese Network is about 20 TB or 3 months of the mobile network USArray represents 500 GB (it depends on the sampling of the recorders).

In addition, the calculation operations downstream may produce even more data than the observation data. To give an order of magnitude, more than 200 TB have to be processed by the Whisper project at the same time. A classical processing produces 8 TB in 5 days. Another computation 'read' 3 or 4 TB and 'produced' 1 TB in 6 hours. Many tests of the signal processing are done and computational data are deleted as and when required.

Nowadays, the earth sciences or more generally, sciences of the universe are widely engaged in data-intensive processing. This leads to design scientific workflow, towards data-intensive discovery and e-Science.

Reflected by the Whisper project, we have to organize the science objectives with the computer constraints. We have to take into account the duration of postdocs and PhD theses, as well as the availability of computer infrastructures and their ease of access. This leads to many questions about software development, including the genericity of computer code and the technical support. But it also influences in terms of choice of appropriate infrastructures.

Even if this project has its own ressources, such a problem of data-intensive processing requires specific tools able to organize distributed data management and access to computational ressources: a data grid environment.

The University of Grenoble offers, thanks to the High Performance Computing (HPC) centre *CIMENT*, this kind of environment with the data management system *iRODS* and the middelware *CiGri*.

It is thanks to the close collaboration between IT ressources of Whisper and the infrastructures of the University that this project has been implemented as we show below.

---

*Cnrs, Isterre, Whisper, Verce, email xav.briand@gmail.com
†CIMENT, Université Joseph Fourier, Technical header of CIMENT
*FP7 ERC Advanced grant 227507, see whisper.obs.ujf-grenoble.fr

2 Software for data-intensive processing

A part of the Whisper project is specificaly dedicated to the IT codes. This includes the design of a specification, an implementation and some optimisations of the sequence of data-management as well as of the computations [†]. This project uses its own IT resources (servers, dedicated bay) as well as the common IT infrastructure of the university. We developed also adaptations for the IT infrastructures and we provide technical support for researchers.

Most of the IT codes are written with the *Python* language and make intensive use of the scientific libraries *Scipy* (fortran and C embeded) and *Obspy* [‡] (essentially the 'read' function) which is dedicated to the seismological community.

The IT codes consist of several tools described schematically in figure 2 and grouped into three parts. The first one concerns the signal processing, the second part permits the computation of the correlations and the last part consists of codes for the analysis of the correlations.

The first package provides a flexible way to process raw data, to specify a pipeline of pre-processing of signal. The user starts by specifying a directory, a set of seismic stations and a set of dates. Then the codes scans the directory and extracts all pieces of seismograms (also called traces) and rearranges them in a specific architecture of files in order to calculate the correlations to the next step. We use here intensively the function 'read' of the library Obspy which allows to open most of the seismogram file formats. The user also defines his own sequence of processings. He can use the functions predefined but also the Python libraries he needs and, moreover, he can add eventually his own functions.

The second package concerns correlations. Roughly speaking, a correlation is an operation with two seismograms (for a given time window) that provides the coherent part of the two seismograms (associated to the 2 stations) which is the seismic waves that propagate between the 2 stations. (Moreover, in some favorables cases, it converges to the Green's function). Thus, the code computes all the correlations and provides an architecture of files that corresponds to all the couples of seismograms (for each date).

Figure 1: Step of the correlations

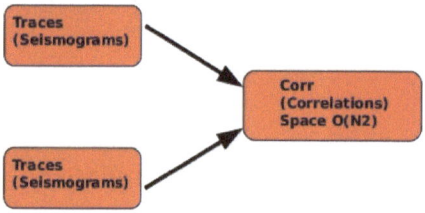

Note that the space complexity is linear for seismograms processing but quadratic for the correlations. We have therefore to store the seismograms processing before compute the correlation in order to benefit of the good complexity. These quadratic space complexity can be critical and lot of effort was made in order to optimize the computation in two direction. First we improve the computation of the fast fourier transform by pre-calculating some "good" combinations of small primes numbers. With this method, we improve of forty percent the time computation in the favorable cases.

Nevertheless, the main optimization was made by testing the behaviour of the carbage collector of Python in order to follow the cache heuristics. More precisely, we do not use the 'gc' module or the 'del' statement but we try to schedule and localize the line of code in order to find the good unfolding that uses the architecture optimally.

The last part of computer codes concerns the analysis of correlations (the virtual new seismograms) with methods such as beamforming, doublet or inversion. We also compute correlations of correlations C3 (also new seismograms). For example, we study the variations in velocity of seismic waves as we illustrate below in figure 9.

These codes permit to process a dataset on a computer laptop. Nevertheless, to take advantage of IT infrastructure at the University of Grenoble, adjustments have been made for the grid computing as we shall see later.

Figure 2: Main sequences of processings of the Whisper Codes

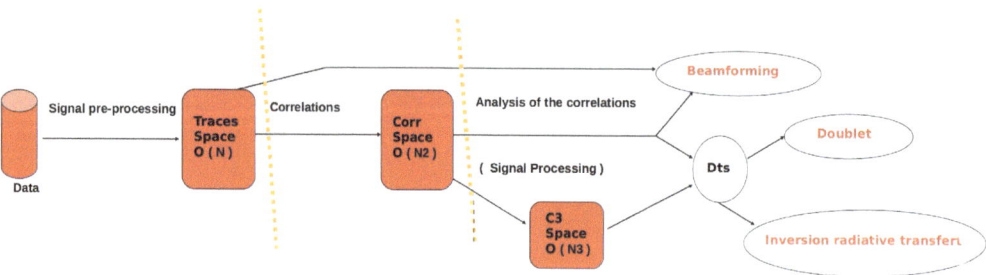

[†] see code-whisper.isterre.fr/html/ (part of the design)
[‡] see www.obspy.org

3 IT infrastructure for grid computing

3.1 CiGri and iRODS synergy

The data-intensive processing needs obviously an IT infrastructure in order to couple storage and computation. In our cases, most of the processings are embarrassingly parallel. The amount of data and the location of available compute nodes suggests the use of a distributed storage system and a grid manager.

The IT infrastructure used here is provided by the *CIMENT* § platform. CIMENT is the High Performance Computing (HPC) center of the Grenoble University. It offers a partial pooling of computing resources (10 computing clusters, 6600 cores and some GPUS) and many documentations for users. Moreover, the computational resources are integrated in a local grid of supercomputers. Associated with a distributed storage, it provides a local data grid environement.

The distributed storage accessible by all the computing nodes of all the clusters is managed by iRODS ¶. Nowadays it represents approximately 700 TB. The grid computing is managed by the *CiGri* ‖ *middleware*, that is part of the OAR ** [4, 7] project (the Resource and Job Management System on which CiGri relies). CiGri and iRODS together build a complementary solution for embarrassingly parallel computations with large input/output distributed data sets.

Furthermore, whith unitary parametric jobs that are reasonnably short in time, CiGri can deal with the best-effort mode provided by OAR. In this mode, grid jobs are scheduled on free resources with a zero priority and may be killed at any time when the local demand of resources increases. This CIMENT organization (independant computing clusters glued together with a best-effort grid middleware and a distributed storage), in place for more than a decade, has proven to be very efficient, allowing near one hundred percent usage of computing resources thanks to small jobs being managed at the grid level.

Furthermore, as the results of the grid jobs are stored into the distributed storage with a unique namespace, iRODS also acts to the user as a centralized controller with a total observation and thus allows the user to monitor its calculation.

3.2 iRODS infrastructure

The Integrated Rule-Oriented Data System (iRODS) is a data managment system offering a single namespace for files that are stored on differents resources that may be on different locations. The administrator can set up rules (microservices) to perform some automatic actions, for exam-

ple the storage of a checksum or an automatic replication to the nearest resource (staging). The user can control himself replications and create user-defined metadata. iRODS exposes a Command Line Interface (the i-commands), an API useable from several programming languages (C, python, PHP,...), a fuse interface, a web gui, and a webdav interface. The meta-catalog is an SQL database, which makes it very efficient for managing additionnal meta-data or making advanced queries (see [6] for an illustration of use). It is not "block-oriented", and thus relies on underlying Posix filesystems. Performance is not the main goal, but when the files are distributed on different resources, the only bottleneck is the meta-catalog (which is centralized).

Figure 3: iCAT and storage nodes

The iRODS storage infrastructure of CIMENT consists of a unique zone with the iCat server and a dozen of nodes as illustrated on figure 3. The nodes are grouped inside 3 different locations, called site A, site B and site C, having heterogeneous WAN connexions. Each site has it's own 10Gbe local network switch.

Figure 4: iRODS resources close to supercomputers

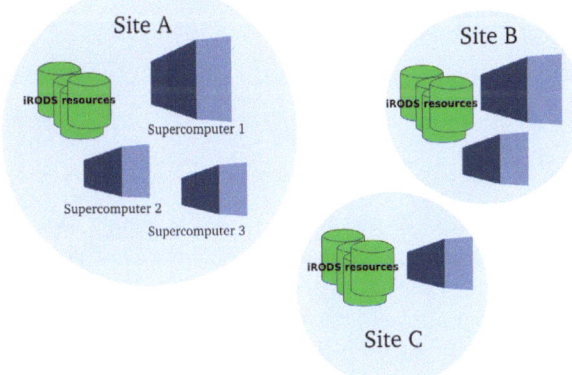

Those 3 sites are located into the 3 main datacenters where the CIMENT supercomputers live, so there are always

---

§ see ciment.ujf-grenoble.fr

¶ see irods.org

‖ see ciment.ujf-grenoble.fr/cigri/dokuwiki

** see oar.imag.fr

close storage resources with the computing nodes (Figure 4). All resources of a given site are groupped into an iRODS resourceGroup and a rule tells that if a data is to be written from a computer of this site, then the data is written by default on a resource randomly chosen inside this group. So, data are always written to a local iRODS resource, using the LAN and not the WAN. Note that site C has only 1Gbe WAN connexions while A and B have 10Gbe WAN connexions (Figure 3). So, to optimize, we've set up automatic staging for site C: when a data is get from site C and the meta-catalog tells that the file is located on a site A or site B resource, then the file is automatically replicated to a resource of site C so that if it is accessed again later, it is not more transfered through the 1Gbe WAN link.

Capacity has now reached 700 TB and is constantly evolving and increases with investment in new projects, as iRODS offers a great scalability by simply adding new storage resources. Each node has currently 2 RAID arrays from 24 to 48 raw TB as illustrated at the figure 5

Figure 5: A storage node

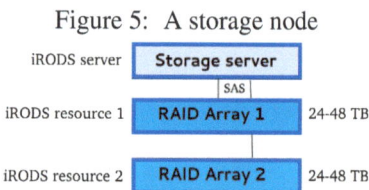

iRODS nodes are running Debian GNU/Linux with Kanif [7] for easy synchronisation of the system administration. CIMENT has set up a web interface where the user can easily check the status of the resources (figure 6).

Figure 6

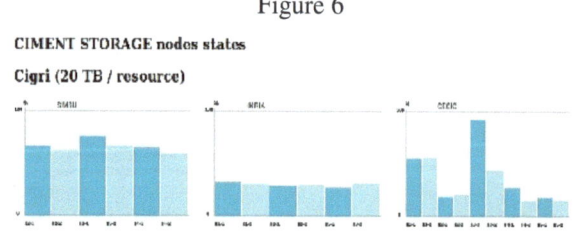

3.3 CiGri infrastructure

The access to 6600 cores of the clusters of the CIMENT platform is achieved through the middleware CiGri. CiGri launches embarrassingly parallel jobs on idle processors of every computing clusters and then optimizes the resources usage which are used for parallel jobs otherwise.

Each cluster of the University of Grenoble uses the resource manager OAR. CiGri acts, among other things, as a metascheduler of OAR. It retrieves the clusters states trough the OAR RESTful API (figure 7) and submits the jobs on free resources without exhausting the local scheduling queues.

Figure 7: Cigri communications

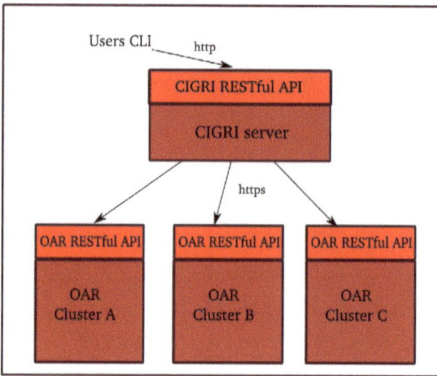

While it may work in normal mode, CiGri is mostly used in best-effort mode and thus provides an automatic resubmission mecanism. CiGri offers a customizable notification system with a smart events management. With those mecanisms, the user can submit a big amount of small jobs, called a campaign, and forget about it until all the jobs of the campaign are terminated or CiGri notifies a serious problem.

Figure 8: Cigri jobs campaign submission

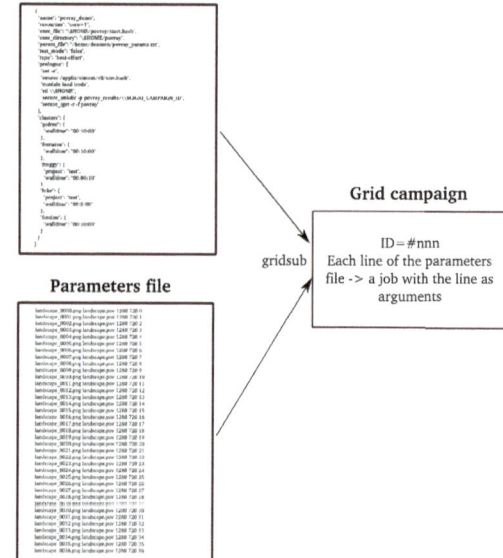

Roughly speaking, in order to run a campaign, the user describes through a file (in the JSON format) the parameters of the campaign such as the accepted cluters, the needed resources, the maximum duration, the location of the codes, a prologue or epilogue script,... Codes and input data are retrieved from iRODS using i-commands into the prologue scripts and the jobs scripts (or using the iRODS API if the jobs are written into a supported language). So, there's no direct connexion between CiGri and iRODS, but the usage is totally complementary through the jobs scripts.

Moreover, the user defines also a file where each line represents a value of the parameter for the user's code. Thus, the number of line of these parameter file corresponds to the number of jobs of the campaign (figure 8).

Users may monitor their campaigns and acts on it whith a CLI or a REST API. Some statistics are provided, such as the completion percentage (in the term of number of terminated jobs), jobs execution rates, automatic re-submissions rate. When a job fails with a non-zero exit status, the user is notified by mail or jabber and requested for an action before submitting further jobs on the same supercomputer: simple acknowledge, acknowledge and re-submission or abort the campaign. Standard and error outputs of the jobs may be easily retrieved from the CiGri host without having to log on the underlying supercomputers. Users may even not be authorized to log-on to a specific supercomputer but allowed to start and manage best-effort jobs on it thanks to CiGri.

CiGri is now at the version 3, which represents a major evolution in terms of modeling and technology (Rest API, Ruby). It is structured around a PostgreSQL database and high level components (Ruby scripting language, Apache with SSL authentication, Sinatra,...).

3.4 Authentication and security

CIMENT has a centralized LDAP infrastructure. Users have the same login on all supercomputers and on the CiGri frontend. As iRODS does not offers a simple and direct LDAP authentication mechanism, we use the simple password method with an automatic synchronisation from our LDAP server to the iRODS database. We also have a script that automatically initializes the iRODS unix environment directly into the users home directory (.irods/.irodsEnv and .irods/.irodsA files) on every supercomputer, so that iRODS authentication becomes completely transparent to the users.

Each site has a filtering router acting as a firewall. As we want all communication schemes to be possible between each iRODS resource (a file might be transfered from a resource to another regardless of the site), we had to open some tcp and udp ports on those firewalls. The range of ports may be defined by the administrator into the iRODS servers configuration file, so that's not an issue.

4 Results and feedback

4.1 Whisper Use Case

Whisper is one of the projects that have made it possible to ensure that the seismic noise brought new observable. This permits to carry out several scientific results including imaging and monitoring. Concerning monitoring, further studies provide news results about sligh variations of seismics waves induced by earthquakes. Many articles are in part due to this project as well as several post-docs and PhD (see [5, 2, 8, 11, 12, 9, 13, 14, 3] and also whisper.obs.ujf-grenoble.fr, rubric publication).

Most of the time, on the computer part, the approach with researchers is as follows. After retrieving data from a data center or directly between persons, we have to assess how this data can be processed. According to the computing time and storage capacity, either we perform operations on a dedicated bay (also host by CIMENT), or either we use the CIMENT infrastructure. It depends also on the ease of computer users and most of the time, at least the last treatments (less computationally expensive) are made locally (Some IT codes are also provided in order to retreive results on distributed storage). For instance, with small datasets (datas from La Reunion or the Alpes) we work only locally. With larger dataset (China, USArray, Japan Network). We use both local and distributed computation.

We focus now on the part of processing that use the data grid environment of CIMENT. But note before that an other aspect, and not least, is the evolution of the specification. Often, students and researchers have new requirements and the IT codes evolves with these specification. We try to be as generic as possible in order to, among other things, to achieve a sufficient level of automatization. However, sometimes we need to develop some parts specifically because of lack of time. These IT problems of specification and development time are among the most complex to evaluate for this type of project.

A first step, for the IT part of the scientific workflow, begins by storing data and also by checking their integrity. We also require that data be in a seismic standard formats. If conversion is necessary, it can be a large data intensive computing and specific codes are developed. To minimize concurrency, we have also to ensure that the data are well distributed on iRODS. Indeed, it happened that the data are too centralized on a resource, so that is truly diminished processing capacity. Some python codes are dedicated to this task and can replicate or spread randomly a Collection from a set of resources to another set.

Let us come back to the package of signal processing of Whisper (the first part of the figure 2). Because each seismogram can be treated separatly it is the simplest case for data grid process. We treated for instance one year of the Japanese seismic Network (HiNet, Tiltmeter and FNet) around of the giant 2011 Tohoku-oki earthquake (6 months before and 6 months after). Note that first we need to convert 9 Terabytes of Japanese Data into around 20 terabytes of a standard format (here mseed or sac). Then we try many processing for the 20 TB (filter, whitening, clipping, ...) and store the results in seismogram with a duration of one day.

As almost all seismological data, the Japanese data are identified by the dates, names of seismic station and subnetworks. Therefore the choice of the modelization, in order to retrieve and distribute the data, follows these seismological metadatas. More precisely, the modelization of the distribution of the computation is made by setting a subset of dates and a subset of stations (This corresponds also to a normal use for researchers that wants to test some processing rapidly with a subset of seismograms). Note also that the distribution is constant with respect to transfer.

In order to get a set of seismograms from the iRODS storage to a computational node, we add Python modules to the Whisper package of seismogram processing. These modules contains classes that permits to retrieve a subset of seismograms (defined by the two subsets described above). More precisely, a first step is either to test directly existence of data or either building of hash tables in order to know the available seismograms. Then we provide an iterator (in this case, a generator for Python) on available data in order to use other methods that performs the i-command 'iget'. The same approach is made for the storage of the results (with the 'iput').

Schematically, the 'Main' Python module of IT code looks like:

```
...
get parameter
...
build the iget commands
perform iget (encapsulation)
...
Codes whisper
for seismogram processing
....
build the iput commands
perform iput (encapsulation)
...
```

Note that these commands of transfer between iRODS and computational nodes can become very difficult to achieve because of the concurrency of the queries. To take into account this obstacle, one develops a module that provides lot of encapsulations of the i-commands (number of try, waiting time, resubmission, with error, 'else' command, etc...). The IT infrastructure provide also wery usefull encapsulations.

We have also to adapt our process for the grid computation with CiGri. We first define a file of parameters 'param.txt' where each line correponds to the parameters of one job on the grid, for instance:

```
cat param.txt

traces160_0_8_0  160 0 8 0
traces160_0_8_1  160 0 8 1
...
traces160_1_8_0  160 1 8 0
traces160_1_8_1  160 1 8 1
...
```

Here we divide the dates in 160 sublists and the stations in 8 sublists. The line "traces160_0_8_1 160 0 8 1" correponds to a job named 'traces160_0_8_1' where we take the sublist of dates of index 0 and the sublist of stations of index 1 (each sublist have the same length +/-1). (There are also other parameters for components and networks not described here.)

For each of the 1280 jobs, a script is launched by CiGri, say 'start.bash', that take for arguments a line of the file parameter. The script, among other things, load necessary library like appropriate python, here 'main.py' and run the code (This is a diagrammatic view).

```
cat start.bash

#!/bin/bash
set -e
...
module load python
...
NumberDate=$2
IndexDate=$3
NumberStation=$4
IndexStation=$5
...
cd DirToCompute/Codes
python main.py NumberDate
        IndexDate NumberStation IndexStation
...
cd
rm -rf DirToCompute
```

Then one defines the json jdl file (job description language), say 'processingSeismogram.jdl'.

```
cat processingSeismogram.jdl

{
"name": "test_processing",
"resources": "core=1",
"exec_file": "$HOME/start.bash",
"param_file": "param.txt",
"type": "best-effort",

"clusters": {

"c1": {
  "prologue": [
  secure_iget -f /iRODSColl/start.bash,
  secure_iget -f /iRODSColl/param.txt,
  mkdir -p DirToCompute,
  secure_iget -rf Codes DirToCompute,
  ... other lines of commands ],
  "project": "whisper",
  "walltime": "00:20:00"
  },

"c2": {
  "prologue": [
  ... lines of commands ],
  "project": "whisper",
  "max_jobs": "450",
  "walltime": "00:30:00"
  },
...
```

```
}
...
}
```

Here the campaign named 'test_processing' take only one core. It run the file 'start.bash' with the parameter file 'param.txt' in mode besteffort. It uses the clusters named 'c1' and 'c2' for the project 'whisper' with the duration defined by 'walltime'. In the prologue, we retrieve script and the parameter file on iRODS as well as the code 'Codes' on iRODS. One can add other parameters for each clusters such that the maximum number of jobs runing at the same time (the variable 'max_jobs').

Concerning the package for the correlations (the second part of the figure 2), we also add similar modules to retrieve on iRODS the subsets of seismograms that have been treated. In order to compute all the correlations that corresponds to all the couple of seismograms we have two types of processes. For a fix subset of dates, either we retrieve one subset of stations and compute all the existing couples for this subset, or either we retrieve two disjoint subsets of stations and compute all the correlations where each seismogram is in a different sublist. This distribution for the computation of the correlations offers a good granularity.

With this kind of distribution for the stations, the transfer bewteen iRODS and the computational node becomes proportional to the distribution, i.e. the number of sublists of stations. Therefore we have to maximize the distribution of the dates and minimize it for the stations in order to obtain a reasonable walltime. Note that it is possible to improve the transfert. However, a fairly simple improvement have to effect that the distribution become dependent on the distributed architecture of computation. In order to keep genericity of the codes we do not change this aspect.

Moreover, special attention was given to selecting the size of files to transfer with iRODS. We do not store each correlation separately, we group them into dictionary to achieve file sizes between 100MB and 500MB most of the time. This order of magnitude seems appropriate for the infrastructure iRODS. For simplicity, we build the grouping according to the parameters of the distribution. This allows to improve the transfer performance. However, unlike a flat architecture, this forced us to develop codes to retrieve the data. Analysis of some subset of correlations (the 3rd step of the figure 2) may request transfer much more than desired.

The mode besteffort also increases the transfer of data because some job are killed and are submitted in an other place. Moreover in some cases, such that a big walltime, we add new steps of iRODS storage for the process in order to store certain intermediate calculation (It is not the case for the correlations because of the good granularity). Note that this may represent a significant development effort.

We focus now on the Japanese computation. With the optimizations described in section 2 and the data-grid environment, the computation of 350 millions individual correlations of the Japanese Network (especially the dense Hi-

Net Network). are done at most a few days. This is a big change that can test many treatments in order to find information in the noise. More precisely the seismogram processing of the Japanese network take half day here (depend obiously on the resampling). Depending on the types of the correlations (stacking, overlap) and also of the availability of the grid (we suppose an usual case here), the computation of all the correlations take between 9h and 20h. The 'iget' i-commands corresponds approximatively to 11TB (with the best-effort mode, some transfers are carried out several times) whereas the 'iput' i-command corresponds here to 3.3TB.

We illustrate with the figure 9, one of the analysis of the correlations that represents an image of change of the velocity of seismic waves in Japan (the giant 2011 Tohoku earthquake, see [3]).

Figure 9

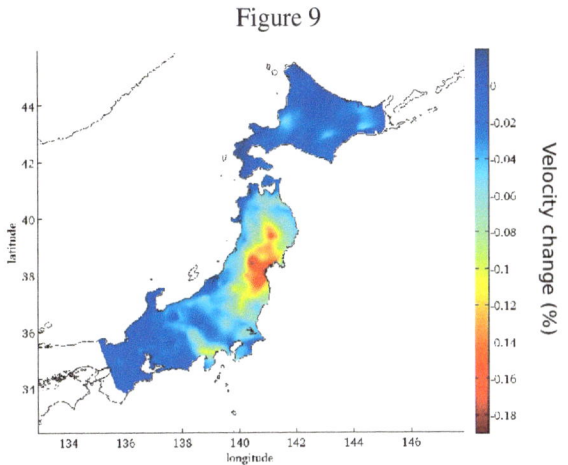

4.2 Infrastructure CIMENT experiences

With the Whisper project, as with some few other projects, we had the opportunity to test and improve the CIMENT IT infrastructure for a data-intensive case. Note also that the data-grid environment is used by many other scientifics projects such that research in Particle Physics [1] or genomics research [10]. For this kind of project, data management and input/output data flows are a big part of the process regarding the actual computing time for the analysis. With a grid of several thousands cpu-cores, such a project may act as a real distributed attack against our storage infrastructures! So, as a first consequence of the deployment of such computing jobs, we had to implement new "limiting" functionnalities into our infrastructure softwares or configurations. For example, CiGri is able to limit the number of submitted jobs for a given campaign on a given cluster if we know that the jobs concurrency can overload our iRODS infrastructure. As another example, iRODS may be configured to limit the number of simultaneous connections. But in this case, the i-command returns with an error and the job may be aborted. There's a "–retry" optionnal argument, but with a high level of concurrency, it increases the load and may completely exhaust the servers with a ton of retry queries. We then developped a wrapper for the i-commands, called secure-i-commands that implements a retry on some specific error cases with an incremental delay to prevent from flooding the iRODS servers. Of course, this can be improved as this leads to a waste of computing resources because they are reserved and not used while waiting. But this is at least a security improvement in case the highler level decision processes does not work as expected because of their high complexity.

Regarding the iRODS choice, it was made on the features: we need a distributed file system suitable for a distributed and heterogeneous grid infrastructure composed of several supercomputers having gateways to reach eachother; and we need a unique namespace. We also need the sites to be independent of the others. It means that a site can be out of service, the data available in another site should still be available. The only case when the entire infrastructure is broken is when the centralized meta-catalog is not reachable. But this part of the infrastructure is located in a highly available datacenter and we may also implement a HA meta-catalog. As it is not posix compliant, iRODS allows more control for the admins and the users. For example, replication can be completely controlled. In our case, we do an automatic replication (staging) of the files when they are got from site A or B to site C, for network optimization. But the user may also want to replicate on every sites to get better performances with big jobs campaigns spread on every clusters of the grid. Also better than a posix filesystem, the users can add custom meta-data to the files and collections allowing them to retrieve data by making some queries with an SQL-like syntax.

Another interesting aspect of iRODS for a project like Whisper is that we can register into iRODS data that are stored on a dedicated posix filesystem. For example, the Japanese raw data were stored on external storage disks that we copied through USB directly onto a server dedicated to the project. We then added this server as an iRODS resource and then registered the data to make them available to the whole grid. In this example, we also used the access rules to set up a read/write policy for the only concerned users.

Open problems and future works

- iRODS has a different transfert policy for "small" files (under 32MB). In this case, files are transiting through the server used by the client instead of a direct connection between the original resource and the client. In a high load context, with a lot of small file transfers, it leads to an overload of the concerned server. A possible solution to that problem could be to share the load among the servers of a given site. For that, we can imagine using a virtual ip address making a round robind on all the ip addresses of the servers of the site.

- We issued some network overloads: automatic staging, as set up for a site that has a lower bandwidth than the others may result in a lot of background iRODS processes doing the replication even when connection rate limit is set up on the servers and even when iget commands are aborted. Under some circumstances, it may completely overload a 1Gb/s ethernet link between 2 sites. Another overload encountered was when the user activates the -Q (use RBUDP (datagram) protocol for the data transfer) option. Even from a single node, this option can cause dramatic overload of network interfaces. Maybe iRODS should implement better rate limit control.

- We made some preliminary tests of the python bindings of the iRODS API (called Pyrods). We noticed that when you have a lot of small operations to do from a python code (like in the Whisper project), either creating a lot of small files, or meta-data, Pyrods may be 10 times faster than a shell loop with i-commands inside. We are now making code examples for the other users to be able to take benefits from this method that would have been useful for the Whisper project.

- We also noticed that a multiprocess program made with python and the PyRODS API improves the speed of a recursive transfer of a collection containing a lot of small files (thousands of 64kb files for example).

5 Conclusion

Despite the problems that have been solved or not, we are currently able to process terabytes of data within a few hours.

Note also that the local data grid (iRODS+CiGri) and more generally the platform of CIMENT as well as the OAR tools permit for the Whisper project to produce significant new results for seismological community in a reduced delay.

Lot of improvements that involve automatization of the processing, and also concerning the scientific workflow have constantly been made. We may say that Whisper+CiGri+iRODS is a great success!

6 Acknowledgements

We thanks Emmanuel Chaljub, the OAR team and the CIMENT team as well as the Whisper project and the Verce project for all their very useful support.

References

[1] C. Biscarat, B. Bzeznik. Synergy between the Ciment tier-2 HPC centre and the HEP community at LPSC in Grenoble (France) *J. Phys.: Conf. Ser. 513 032008, 2014*

[2] P Boué, P Poli, M Campillo, H Pedersen, X Briand, P Roux. Teleseismic correlations of ambient seismic noise for deep global imaging of the Earth. *Geophysical Journal International 194 (2), 844-848.*

[3] Brenguier, F., M. Campillo, T. Takeda, Y. Aoki, N. M. Shapiro, X. Briand, K. Emoto, H. Miyake (2014). Mapping pressurized volcanic fluids from induced crustal seismic velocity drops. *Science, 345, 80-82, 2014.*

[4] Nicolas Capit, Georges Da Costa, Yiannis Georgiou, Guillaume Huard, Cyrille Martin, Grégory Mounié, Pierre Neyron, Olivier Richard, A batch scheduler with high level components. *Cluster computing and Grid 2005 (CCGrid05), Cardiff, Royaume-Uni, 2005.*

[5] JH Chen, B Froment, QY Liu, M Campillo. Distribution of seismic wave speed changes associated with the 12 May 2008 Mw 7.9 Wenchuan earthquake. *Geophysical Research Letters 37 (18).*

[6] Gen-Tao Chiang, Peter Clapham, Guoying Qi, Kevin Sale and Guy Coates, Implementing a genomic data management system using iRODS in the Wellcome Trust Sanger Institute. *BMC Bioinformatics,12(1):361+, September 2011.*

[7] Benoît Claudel, Guillaume Huard, Olivier Richard. TakTuk, Adaptive Deployment of Remote Executions. *Proceedings of the International Symposium on High Performance Distributed Computing (HPDC) (2009) 91-100.*

[8] G. Hillers, M. Campillo, K.-F. Ma. Seismic velocity variations at TCDP are controlled by MJO driven precipitation pattern and high fluid discharge properties. *Earth and Planetary Science Letters, Volume 391, 1 April 2014, Pages 121-127.*

[9] Macquet M., Paul A., Pedersen H., Villase nor A., Chevrot S., Sylvander M. and the PYROPE working group. Ambient noise tomography of the Pyrenees and surrounding regions : inversion for a 3-D Vs model in a very heterogeneous crust. *accepted in Geophysical Journal International.*

[10] Münkemüller T., Lavergne S., Bzeznik B., Dray S., Jombart T., Schiffers K., Thuiller W. How to measure and test phylogenetic signal. Methods in Ecology and Evolution, 3: 743-756, 2012.

[11] Obermann, A., T. Planès, E. Larose, and M. Campillo. Imaging preeruptive and coeruptive structural and mechanical changes of a volcano with ambient seismic noise. *J. Geophys. Res. Solid Earth, 118, 6285-6294.*

[12] P. Poli, M. Campillo H. A. Pedersen, Body-wave imaging of Earth's mantle discontinuities from ambient seismic noise. *Science 338 (6110), 1063-1065.*

[13] D. Rivet, M. Campillo, M. Radiguet, D. Zigone, V. Cruz-Atienza, N. M. Shapiro, V. Kostoglodov, N. Cotte, G. Cougoulat, A. Walpersdorf and E. Daub Seismic velocity changes, strain rate and non-volcanic tremors during the 2009-2010 slow slip event in Guerrero, Mexico. *Geophys. J. Int. 2014;196:447-460.*

[14] Rivet D., Brenguier F., Clarke D., Shapiro N.M., Peltier A. Long-term dynamics of Piton de la Fournaise volcano from 13 years of seismic velocity changes measurements and GPS observation. *J. of Geophys. Res. 2014; submitted.*

Development of a Native Cross-Platform iRODS GUI Client

Ilari Korhonen, Miika Nurminen
IT Services, University of Jyväskylä
PO Box 35, 40014 University of Jyväskylä, Finland
ilari.korhonen@jyu.fi, miika.nurminen@jyu.fi

ABSTRACT

This paper describes activities on the research IT infrastructure development project at the University of Jyväskylä. The main contribution is a cross-platform iRODS client application with a rich graphical user interface. The client application is fully native and builds from a single C++ codebase on all of the platforms on which iRODS 4.0 is supported. The application has a responsive UI with native look & feel and enables drag & drop integration to the desktop. This is made possible by basing the development of the client application on top of the Qt 5 framework and an object-oriented C++ framework for iRODS which is being developed with the client application. The object-oriented framework wraps around the native iRODS 4.0 C/C++ client API library and provides object-oriented interfaces to iRODS protocol operations e.g. a fully object-oriented iRODS General Query (GenQuery) interface used by the client application has been implemented in this C++ framework. By developing on top of the native C/C++ iRODS API library, the plugin architecture of iRODS 4.0 can be fully leveraged in authentication (e.g. Kerberos) and network transport (e.g. SSL) modules without any additional complexity.

Keywords

Research data, metadata management, infrastructure, iRODS, client software, graphical user interface.

INTRODUCTION

There is an increasing demand for IT services for researchers that span the full "stack" of storage and computation infrastructure for research data with metadata support and widely available interfaces for information extraction and reporting. On one hand, the scope of computationally intensive datasets currently spans all fields of research necessitating university-wide support for scientific computing and research data management. On the other hand, funders and institutions (e.g. EU's Horizon 2020) have expressed an increased demand for opening all research materials related to publicly funded research. This can be seen as a continuation of recent development with institutional repositories (e.g. DSpace, EPrints, and Fedora) supporting the "green" way of Open Access for publications.

In Finland, the National Research Data Initiative (TTA) and Open Science and Research (ATT) projects have developed research data infrastructure and promoted open access. The Ministry of Education and Culture is considering to include openness (at least for publications) as an element in the funding model for the Finnish universities [1], pushing the need to get (meta)data from research data as well. In this paper, we describe research IT infrastructure development at the University of Jyväskylä, focusing on iRODS. An iRODS client application is introduced and briefly evaluated with respect to planned data management processes. Finally, prospects for development are outlined.

RESEARCH IT INFRASTRUCTURE DEVELOPMENT AT THE UNIVERSITY OF JYVÄSKYLÄ

In this paper, we describe some of the recent development efforts related to research IT infrastructure at the University of Jyväskylä. The discussion is focused on the project codenamed "Kanki" (=meaning e.g. in Finnish "a rod" and in Japanese "cold" or "frost") – a native cross-platform iRODS client application with a rich graphical user interface

iRODS UGM 2015 June 10-11, 2015, Chapel Hill, NC

based on Qt[1] framework. The client application is targeted towards researchers of various disciplines as well as other interest groups utilizing or curating research data (e.g. librarians), possibly lacking the expertise to use the iRODS command-line interface. Our client application will enable the users to utilize the full power of an iRODS data grid complete with powerful data management functions such as schema management and validation of metadata.

Background

There is a relatively long tradition related to the advancement of open access and research data management at the University of Jyväskylä. Theses were published online by the Jyväskylä University Library as early as 1997, leading to the introduction of the DSpace-based institutional repository JYX[2] in 2008 [2]. A university-wide working group for parallel publishing and administration of research material was commenced in 2009, resulting in the development of a mechanism for parallel publishing of publication files from the research information system TUTKA[3] to the institutional repository. Even though the working group had identified various types of research materials that should be preserved, support for managing research data in a standardized way (considering both tools and data-related processes) was incoherent, differing between research groups. It proved to be of considerable difficulty to advance standard data management practices when the research itself is done independently of administrative processes, often using specialized tools and software for e.g. analysis or other parallel computation on datasets. Many older research materials are still in analogue form – and even those in digital form are often stored in either removable media, portable hard drives or in the best case – file servers. Some faculty-specific solutions such as YouData[4] are in use, but most datasets lack standardized metadata descriptions, complicating data discovery and reuse.

Recently, University of Jyväskylä has taken an active approach on managing research data and infrastructures. In September 2014, JYU was the first Finnish university to have published its official principles for research data management [3]. The development project for research IT infrastructure and research data management has been active in 2013-2015. Project activities include the adoption and integration of the Dataverse Network[5], the development of a university-wide iRODS grid infrastructure for research data storage, and the surveying of essential datasets in the faculties for which to develop iRODS data management services. The iRODS platform has been selected as the primary focus of development activities in the project. Overall architecture is based on separated responsibilities between the systems. Even though some institutional repositories have been augmented for publishing research data, support for managing the data during the whole research life cycle is inadequate (unless extensively customized, which can be a problem from maintenance point of view). The metadata used in research datasets has considerably more variation compared to metadata typically used in the repositories (e.g. Dublin Core). Providing access to data in repositories is no longer enough since people want to *do things* with that data [4]. iRODS responds to this need internally. Dataverse has potential to respond to the external needs with citable datasets and analysis functionality.

The iRODS-related development activities at JYU include the development of a secure, scalable, high-performance, high-availability iRODS data grid infrastructure for university-wide deployment with infastructure automation, the development of server-side iRODS modules for e.g. metadata autoextraction and data anonymization, and finally, the development of a native cross-platform iRODS GUI client to enable schema-based metadata management with validation capabilities and to serve as a platform for future iRODS-based applications. A common metadata model for JYU research data management is being developed with JYU Library, based on national specifications. The Finnish national research data storage service IDA[6] – maintained by the Finnish IT Center for Science CSC – is built on top of iRODS as well. Collaboration with the IDA development has been planned. The overall goal of the project is in the advancement of the IT service culture to improve the acceptance of centralized services among the researchers and to be able to provide added value compared to isolated legacy solutions.

---

[1] http://www.qt.io/
[2] https://jyx.jyu.fi/
[3] http://tutka.jyu.fi/
[4] http://youdata.it.jyu.fi/
[5] https://dvn.jyu.fi/
[6] https://www.tdata.fi/en/ida

Infrastructure Development

The IT infrastructure at the University of Jyväskylä is a largely consolidated one with most of its servers residing as virtual machines in a VMware vSphere 5 cluster. Separate physical servers are being used alongside virtualization for performance critical computing or I/O-intensive applications while strongly favouring virtual servers. Shared storage is being provided by EMC VNX Series SAN/NAS unified storage arrays connected to Fibre Channel and 10 Gbps Converged Enhanced Ethernet fabrics. After initial evaluation and testing of iRODS it was promptly concluded that a single deployment of an iRODS iCAT server is insufficient to provide the performance targeted for scalable use and would not be highly available without the use of (performance limiting) hardware virtualization. This prompted the design of a scalable and inherently highly available infrastructure for iRODS iCAT deployment at JYU.

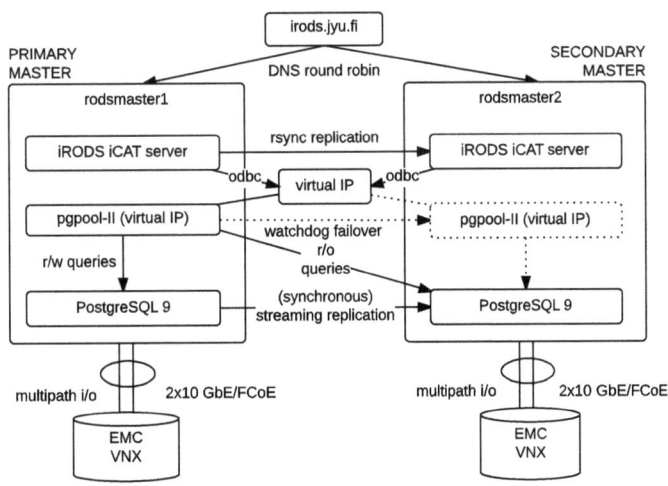

Figure 1. Illustration of the scalable HA model for JYU iRODS deployment.

A critical point from performance point of view and simultaneously a *single point of failure* is the database server used to host the iRODS iCAT database. For the database instance to be able to both scale out from a single server, and to withstand a loss of a server without compromising the integrity of the database – a properly clustered database solution is a necessity. An Oracle RAC database high-availability cluster solution wasn't considered feasible because of the prohibitive pricing per processor core of Oracle database server products. After evaluating the possible alternatives (e.g. HAIRS [5]) for a high-availability load balanced iRODS iCAT cluster we propose the following model for a scalable iRODS iCAT deployment (Figure 1). The solution is built on a highly available pooled configuration of a PostgreSQL 9 database, using PostgreSQL 9 streaming replication and PgPool-II for load balancing and HA failover on top. The system can be built on configurations with at least two servers and scales horizontally on read-only transaction performance and is able to withstand the loss of a single server and having no single point of failure.

On the very lowest level of the configuration are the PostgreSQL 9 database server instances, one initially set up as the primary database master, the other ones as read-only database replicas, which receive streaming replication from the master database. On top of the PostgreSQL 9 database server instances is PgPool-II, which is configured to a HA configuration via the built-in watchdog facility. PgPool-II is configured on both or all of the iCAT servers to *share* an IP address (in an HA subnet available in a private VLAN shown to the iRODS iCAT servers) such that if the current master PgPool-II host goes down there will be an *escalation* procedure to select a new PgPool-II master host which will take over the *virtual* PostgreSQL pool IP address in the HA subnet. The high-availability PgPool-II configuration – resident in all of the iRODS iCAT servers connected via the HA subnet – is aware of the state and health of all of the underying database servers and in the case of a failover event (master PostgreSQL database backend health check failed) executes a *recovery* operation to an available PostgreSQL 9 hot standby server, which becomes the new master of the database as ordered by PgPool-II and starts accepting read-write transactions to the database. The recovery process sets the other server(s) as read-only hot standby replicas of the new master database.

The iRODS iCAT server(s) are set up on top of the PgPool-II managed database cluster such that the iCAT server(s) are set to use a pooled database available at the pool virtual IP. This way the iRODS iCAT may utilize the entire database cluster for increased performance, since the PgPool-II middleware load balances the read-only transactions throughout the cluster. Read-write transactions are sent only to the master database and synced to the other nodes with streaming replication. Some load balancing of the iRODS connections can be accomplished using a simple DNS round-robin set up with the caveat that iRODS itself is not DNS round-robin aware. Additionally the DNS name service caching in use in common operating systems hinders with the round-robin of resolved IP addresses. We hope that in the future this aspect would be addressed in the development of iRODS. For the time being this solution provides some load balancing between clients of iRODS connections to two or more iRODS iCAT servers.

The deployment of new iRODS instances to a HA pair of servers is being done at JYU with Ansible infrastructure automation. The setup has been parametrized such that a new iRODS instance can be specified simply by Ansible *group variables* for the group of iRODS iCAT servers and *host variables* to specify the hosts themselves. This is believed to be useful not only for quick deployment of development servers but also for deploying new iRODS instances for special use cases. For example, a specialized instance of iRODS deployment is in planning stage for the new Jyväskylä Center for Interdisciplinary Brain Research[7].

PROJECT KANKI – A NATIVE CROSS-PLATFORM GUI CLIENT APPLICATION FOR IRODS

During the research IT infrastructure development project several needs have risen for iRODS-based research data (and *meta*data) management. These prompted the need for iRODS user interface development. The utmost important of these was the need for secure data and metadata transfer. Other specific needs not properly accomodated by other existing freely available solutions included the graphical iRODS search tool with arbitrary search criteria formation for data discovery, metadata schema management with visual namespace and attribute views and readiness for metadata schema validation for data quality assurance. Some open source projects provided these features partially, but other projects such as Davis[8] were discontinued and thus rendered unsupported after the introduction of iRODS 4.0.

About the Development

To implement these user interface features for iRODS-based research data management at JYU, a software project was started – eventually codenamed "Kanki" – to build an iRODS 4.0 compatible client application with integration to Kerberos authentication, the option to use iRODS 4.0 SSL secured connections, to develop extensible data and metadata management features, and to serve as a framework for iRODS integration to different kinds of scientific software. The introduction of iRODS 4.0 and the incorporation of the modular architecture in both iRODS 4.0 server and client side made the native iRODS client library more attractive for client-side development than any of the other options available. The possibility to use e.g. Kerberos authentication or SSL transports in the iRODS connections out-of-the-box – without having to resort to e.g. IPSec for transport layer security – made a convincing case for C++ to be used also for client-side development instead of more popular alternatives like Java or Python.

Since the goal of the development was to produce a cross-platform application while still remaining fully native, the widely adopted Qt framework proved to be an excellent choice as a development platform. At the University of Jyväskylä we deploy all of the major platforms i.e. Linux, Mac OS X and Windows, with Red Hat Enterprise Linux being the prominent Linux distribution with a campus license. For developing a cross-platform application targeted to all these platforms, the Qt framework provides exceptional support for compiling from a single codebase.

Challenges

A working build configuration for Mac OS X took some effort since the newer versions of OS X, namely versions later than OS X 10.8 Lion caused some difficulties for building iRODS. The version of the boost libraries compatible with iRODS 4.0 proved to be incompatible with the OS X provided clang compiler. GCC 4.8 or later proved to be a

---

[7] http://cibr.jyu.fi/

[8] https://code.google.com/p/webdavis/

working solution. There were other issues as well caused by a dynamic linker symbol conflict with OS X bundled MD5 library functions and ones provided with iRODS. The most severe consequence of this seemed to be the inability to use the native auth module in OS X builds of iRODS 4.0, since the runtime dynamic linker resolved some of the MD5 symbols to the iRODS bundled ones and others to the OS X provided ones, causing the corruption of the memory buffer used to compute the MD5 challenge response. This was first worked around by not using native authentication – which is not to be used at JYU iRODS at all anyway. A workaround solution was found for the issue by changing linkage of the auth module in a way which resolves the symbol conflict. Currently, this issue has been resolved in the iRODS master branch.

Windows still remains as an unsupported platform since iRODS 4.0 isn't Windows compatible at the time of writing of this paper. With Windows support added to the iRODS codebase our client can be built on Windows as well.

Features

The client (see Figure 2) is intended to eventually serve as a bona fide alternative user interface to iRODS icommands – the reference user interface for iRODS. Implementing all of this functionality in a native "desktop" application will enable the users to harness the full power of iRODS with native application performance and the usability of a graphical user interface.

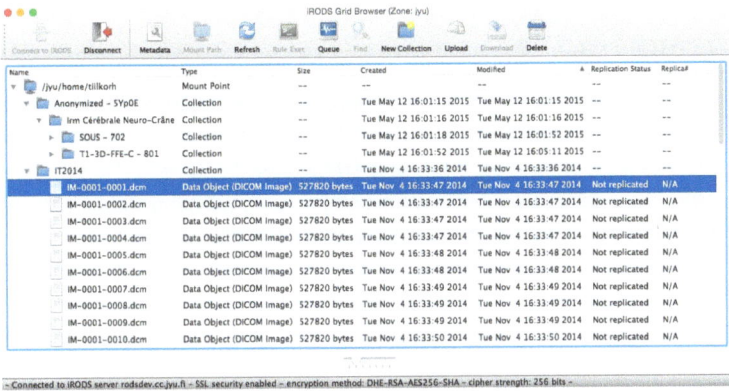

Figure 2. iRODS Grid Browser window in the client application.

Additionally, the client has some specialized features for metadata management not found in currently available iRODS clients. Metadata schema management has been implemented with features like namespace and attribute management. Below is an example of attribute descriptions in the XML metadata schema configuration.

```
<irods:namespace prefix="fi.jyu.irods." label="University of Jyväskylä">
  <irods:attribute name="metadata.modified" unit="false" editable="false">
    <irods:label>Metadata Modification Time</irods:label>
    <irods:displayFilter type="regExp"><irods:regExpRule>(\d+)-(\d+)-(\d+).(\d+):(\d+):(\d+)</irods:regExpRule>
    <irods:regExpFilter>\3.\2.\1 \4:\5:\6</irods:regExpFilter></irods:displayFilter>
  </irods:attribute>
  <irods:attribute name="language" unit="false" editable="true">
    <irods:label>Language</irods:label><irods:values strict="true">
    <irods:value name="ISO6392:FIN"><irods:label>Finnish</irods:label></irods:value>
    <irods:value name="ISO6392:ENG"><irods:label>English</irods:label></irods:value>
    <irods:defaultValue>ISO6392:ENG</irods:defaultValue>
  </irods:values></irods:attribute>
</irods:namespace>
```

In the metadata schema configuration namespaces are identified along with attributes defined in the namespaces. Namespaces and attributes can be defined having labels for the ease of use of the metadata editor. Additionally, attributes can be defined with "display filters" which transform the attribute value stored in the iCAT database to a more human-readable form. Currently only regular expression display filters are implemented but others are planned such as a GenQuery filter for translating an attribute with a GenQuery and a JSON filter for transforming JSON encoded attributes into a more visual form for display purposes. A validator interface is planned for the metadata editor enabling client-side validation of metadata entries to the iCAT database e.g with the defintion of allowed values for the attribute in the metadata schema the editor would present the user with a drop-down list of acceptable values. Figure 3 views the metadata manager as currently implemented.

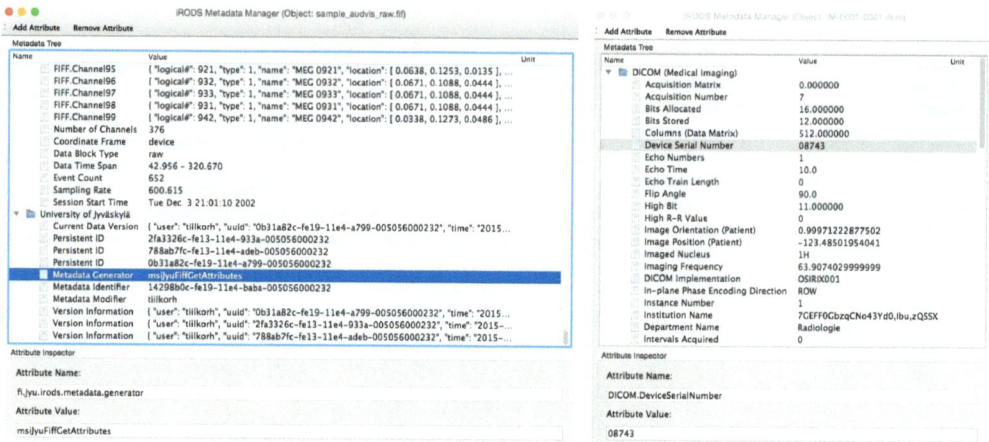

Figure 3. iRODS Metadata Manager windows in the client application.

At the core of the client application is an object-oriented interface for iRODS which wraps around the native C/C++ iRODS API. It is planned that these iRODS specific interface classes are to form an object-oriented C++ iRODS framework. Below is an example of building and executing a GenQuery for retrieving metadata AVU triplets for either a data object or a collection. The code sample is taken from the metadata model class of the client application.

```
Kanki::RodsGenQuery metaQuery(this->conn);
int status = 0;
if (this->objDatum->objType == DATA_OBJ_T) {
    metaQuery.addQueryAttribute(COL_META_DATA_ATTR_NAME);
    metaQuery.addQueryAttribute(COL_META_DATA_ATTR_VALUE);
    metaQuery.addQueryAttribute(COL_META_DATA_ATTR_UNITS);
}
else if (this->objDatum->objType == COLL_OBJ_T) {
    metaQuery.addQueryAttribute(COL_META_COLL_ATTR_NAME);
    metaQuery.addQueryAttribute(COL_META_COLL_ATTR_VALUE);
    metaQuery.addQueryAttribute(COL_META_COLL_ATTR_UNITS);
}
// add a query condition for object name
metaQuery.addQueryCondition(this->objDatum->objType == DATA_OBJ_T ? COL_DATA_NAME : COL_COLL_NAME,
                        Kanki::RodsGenQuery::isEqual, this->objDatum->objName);
// if we are querying a data object also specify collection path
if (this->objDatum->objType == DATA_OBJ_T)
    metaQuery.addQueryCondition(COL_COLL_NAME, Kanki::RodsGenQuery::isEqual, this->objDatum->collPath);
// execute genquery and get status code from iRODS API
if ((status = metaQuery.execute()) < 0) {
    // error reporting code
}
else {
    std::vector<std::string> names = metaQuery.getResultSet(0);
    std::vector<std::string> values = metaQuery.getResultSet(1);
    std::vector<std::string> units = metaQuery.getResultSet(2);
}
```

DISCUSSION

Our main concern with iRODS is related to metadata management – both in terms of metadata quality assurance and the scope of the supported metadata structures. It has been observed that in general, at least 5% of the information present in manually created databases is erroneous [6]. Lessons learned in the institutional repositories domain from self-archiving of publications should be taken into account [7] - researchers should not be responsible for filling metadata fields alone, but as a collaborative process assisted by librarians. On one hand, it is important to allow researchers to edit the metadata entered into the system to get the first-hand insight to the datasets, but it is up to the librarians to ensure that the metadata is in a consistent form, and, if necessary, to "clean" the metadata afterwards. An essential requirement is that the metadata resides in a centralized repository such that two-way synchronizations (and in particular duplicate manual entries) are kept at minimum (i.e. master data is managed) and metadata is reused when possible. Features of our iRODS infrastructure facilitate this goal by preserving information about the latest metadata update (user, timestamp and the metadata UUID), extensible validation functionality, and cascading collection-level metadata to data objects. Practices that yet need to be implemented include duplicate detection, metadata batch editing, and delete/replace on list-like metadata – an effective way to clean up records with misspellings, but to be used with caution [8].

Scoping the supported metadata structures is related to metadata quality. If one is confined to a standard minimum metadata set, object-specific metadata can be perceived as a small set of plain-text fields, resulting to little attention in metadata validation. This alone can be a quality problem if conventions are not followed – especially if data is aggregated from multiple sources [7]. However, depending on the domain, essential metadata may be much more involved, containing diverse compound fields (e.g. MARC in the library domain), or multiple entity classes that may refer to each other. An example is the domain of cultural heritage, where a trend of shifting from item-centric cataloging (physical objects as the primary entities) to event-centric documentation (e.g. CIDOC CRM – events related to the objects) is taking place [9]. Individual fields may be too coarse-grained to represent events, but compound fields specified in a JSON-like structure may be part of the solution. For selected fields, utilization of ontologies (e.g. controlled keywords in the Finnish national Finto service) and other external sources (e.g. name disambiguation with ORCID identifiers) becomes relevant, but needs additional development. One prospect might be declaring special data objects *contained in* iRODS as internally controlled authorities for recurrent, shared metadata (represented as a look-up list), akin to the solution applied in DSpace-CRIS[9]: entities function as authorities for item metadata [10].

As a development framework, Qt provides several benefits compared to both OS-specific (i.e. iExplore [11]) or even web-based solutions. Unified look and feel could be accomplished with other languages or frameworks, but C++ -based compilation and direct linking to iRODS and system libraries provides the best possible performance. Web clients such as iDrop[10] contain useful functionality from data access perspective (no need to install additional software) and we except them to be used for some use cases. However, we argue that a purely web-based client is insufficient for more involved data management (i.e. performing computation on multiple versions of the dataset with close integration to local filesystem). iDrop does not provide schema-specific validation which is a problem from data quality perspective. Another problem with web-based interfaces is – despite recent HTML5 improvements – that file management cannot be implemented with explorer -like capabilities. Even though it is possible to implement drag'n'drop support from local filesystem to browser, it would be limited to file transfer. Mass edit functions, versioning, or 2-way synchronization between web-based view and local filesystem would need an additional plug-in or client application. Therefore, a cross-platform native client application is a critical factor to improve the utilization rate of the system since most users *already expect* a user experience as streamlined (but lacking in metadata, validation, or security aspects) as in popular cloud-based file-sharing services such as Dropbox or Google Drive.

---

[9] http://cineca.github.io/dspace-cris/

[10] https://github.com/DICE-UNC/idrop

CONCLUSION

Future prospects with our IT infrastructure development project include increased integration – both on the storage and application levels. JYU iRODS could be used as a general purpose storage middleware beyond research data for e.g. the JYU digitization center with metadata imported from the book scanner workflow. The institutional repository JYX would also benefit from iRODS storage in contrast to typical filesystem-based asset store. Potential application-level integrations include connections from iRODS to Dataverse for publishing datasets, and the Current Research Information System (CRIS – currently in procurement at JYU [12]) . Whereas an institutional repository, iRODS or Dataverse are used to store the *outputs* of a research project, a CRIS system provides information about the research projects themselves. This includes metadata related to publications and project-related documentation (e.g. research plans, funding decisions). A CRIS could be used as a data hub combining information about research infrastructures, projects, and outputs – used for aggregating metadata from other sources (i.e. bibliographic databases) as well as feeding it to other systems, such as a data warehouse, an institutional repository, or an iRODS grid.

The development of our client is still at early stages. The iRODS infrastructure and our client have been presented to other Finnish HEIs and the National IT Center for Science CSC at the National IT Days for Higher Education (*IT-päivät* in Finnish) and other occasions. Our solution has provoked interest, showing potential to be of use in other universities as well as IT services operated by CSC. We intend the development process to be a collaborative effort and plan to publish the code under an open source licence. We welcome suggestions regarding the features for the UI, validation, and data description format and hope that the software will be utilized in other institutions.

ACKNOWLEDGMENTS

The authors thank the Head of IT Management at the University of Jyväskylä, D.Sc. Antti Auer, for his contribution on research data management policies at JYU, enabling management support for the development project.

REFERENCES

[1] A. Neuvonen *et al.*, "ATT-vaikuttavuusselvitysryhmän raportti [ATT effectiveness working group report]," Tech. Rep., 2015. [Online]. Available: https://confluence.csc.fi/pages/viewpage.action?pageId=45389145

[2] P. Olsbo, "Institutional repository as a center of services," 2012, poster at IFLA World Library and Information Congress. [Online]. Available: http://www.libraries.fi/en-GB/finnishlibraryworld/presentations/

[3] A. Auer and S.-L. Korppi-Tommola, "Principles for research data management at the University of Jyväskylä," Tech. Rep., 2014. [Online]. Available: https://www.jyu.fi/tutkimus/tutkimusaineistot/rdmenpdf

[4] T. Walters, "Assimilating digital repositories into the active research process," in *Research Data Management – Practical Strategies for Information Professionals*, J. M. Ray, Ed. Purdue University Press, 2014.

[5] Y. Kawai and A. Hasan, "High availability iRODS system (HAIRS)," in *Proceedings of iRODS User Group Meeting 2010*. Data Intensive Cyberinfrastructure Foundation, 2010.

[6] A. van den Bosch, M. van Erp, and C. Sporleder, "Making a clean sweep of cultural heritage," *IEEE Intelligent Systems*, vol. 24, no. 2, pp. 54–63, 2009.

[7] J. W. Chapman, D. Reynolds, and S. A. Shreeves, "Repository metadata: Approaches and challenges," *Cataloging & Classification Quarterly*, vol. 47, no. 3-4, pp. 309–325, 2009.

[8] M. Nurminen and A. Heimbürger, "Representation and retrieval of uncertain temporal information in museum databases." in *Information Modelling and Knowledge Bases XXIII*. IOS Press, 2012.

[9] A. Häyrinen, "Open sourcing digital heritage – digital surrogates, museums and knowledge management in the age of open networks," Ph.D. dissertation, University of Jyväskylä, 2012.

[10] S. Mornati and A. Bollini, "DSpace-CRIS: an open source solution," 2013, presented at euroCRIS Membership Meeting. [Online]. Available: http://www.eurocris.up.pt/?page_id=13

[11] B. Zhu, "iExplore for iRODS distributed data management," in *Proceedings of iRODS User Group Meeting 2010*. Data Intensive Cyberinfrastructure Foundation, 2010.

[12] M. Nurminen, "Preparing for CRIS: Challenges and opportunities for systems integration at Finnish universities," 2014, poster at Open Repositories 2014.

Pluggable Rule Engine Architecture

Hao Xu
DICE Center
University of North
Carolina at Chapel Hill,
NC 27599, USA
xuh@email.unc.edu

Jason Coposky
Renaissance Computing
Institute (RENCI)
100 Europa Drive Suite
540 Chapel Hill, North
Carolina 27517
jasonc@renci.org

Ben Keller
Renaissance Computing
Institute (RENCI)
100 Europa Drive Suite
540 Chapel Hill, North
Carolina 27517
kellerb@renci.org

Terrell Russell
Renaissance Computing
Institute (RENCI)
100 Europa Drive Suite
540 Chapel Hill, North
Carolina 27517
unc@terrellrussell.com

ABSTRACT

We describe a new development in the next release of iRODS. The pluggable rule engine architecture allows us to easily create new rule engines as plugins and run multiple rule engines concurrently. The pluggable rule engine architecture allows easy implementation and maintenance of rule engine plugin code and offers significant performance gains in some use cases. The pluggable rule engine architecture enables modular incorporation of features from other programming languages, allows efficient auditing of interactions between user-defined rules and the iRODS system, and supports full interoperability between rules and libraries written in different languages. This design allows us to easily incorporate libraries designed for different programming languages, for example, Python, C++, etc., into the policy sets, significantly enhancing the capabilities of iRODS without syntactic overhead. This new architecture enables a wide range of important applications including auditing, indexing, and modular distribution of policies. We demonstrate how to create the Python rule engine plugin and how to create user defined policy plugins.

Keywords

Pluggable Policy, Rule Engine, Plugin Architecture

INTRODUCTION

In this paper, we are going to describe a new development in the next release of iRODS. The pluggable rule engine architecture allows us to easily create new rule engines as plugins and run multiple rule engines concurrently. The pluggable rule engine architecture allows easy implementation and maintenance of rule engine plugin code and offers significant performance gains in some use cases. The pluggable rule engine architecture enables modular incorporation of features from other programming languages, allows efficient auditing of interactions of user-defined rules and the iRODS system, and supports full interoperability between rules and libraries written in different languages. This design allows us to easily incorporate libraries designed for different programming languages, for example, Python, C++, etc., into the policy sets, significantly enhancing the capabilities of iRODS without syntactic overhead. This new architecture enables a wide range of important applications including auditing, indexing, and modular distribution of policies.

Users of iRODS have expressed the following areas of improvement:

iRODS UGM 2015 June 10-11, 2015, Chapel Hill, NC

- customization of error handling in pre and post PEPs.

- calling microservices written in other languages directly.

- native performance for event tracking rules.

- modular distribution of policies.

- full auditing of data access operations.

- reduce manual change when upgrading.

- new policy enforcement points.

The pluggable rule engine architecture addresses these challenges.

THE DESIGN

In this section, we overview the key designs in the pluggable rule engine architecture.

iRODS Features

iRODS supports a wide range of plugin types. This allows the core iRODS to be independent from the components that it uses. For example, the database plugin allows iRODS to use different databases without changing the core code. Each plugin has a set of defined operations that it has to provide. The core interacts with plugins only through those operations. One benefit of this design is that we can easily capture all state changing operations by looking at plugin operations. And we can show that such capture is complete in the following sense. If we want to capture all database operations, we only need to look at database plugin operations. Because of the ignorance of the underlying implementation of these operations, the core cannot perform any additional operations than those provided by the plugin architecture. Therefore, if we capture all operations in the plugin architecture, we capture all state changing operations.

In iRODS, a pair of pre and post PEPs are automatically generated for every defined plugin operation. This way we ensure that all policy enforcement points are present. Having the capability to write policies for every state changing operation, we make the complete information about each operation available to the PEPs by making the argument and environment in which the operation is called available to the PEPs. This way the PEPs can determine what to do based on this information.

Formally speaking, let Op denote the set of plugin operations, and Act denote the set of actions, with

$$Op \subset Act$$

Let f denote the function that generates an action from a plugin operation. For example, given a plugin operation, the plugin architecture generates a PEP-added action Act comprising of pre and post operations PEPs as follows:

$$f : Op \to Act$$

$$f[op(args, env)] = pre_{op}(args, env); op(args, env); post_{op}(args, env)$$

Here the sequential combination operator can be thought of as the monadic *bind* operator. This formalism can be used to adopt a wide-range of applications. One of the disadvantages of this design is that the semantics of f must be fixed in an iRODS implementation, for example, how the error is handled. And the particular form f lacks principal error handling semantics, i.e., one which fits all of our users' use cases by just varying pre and post PEPs. For example, should we make op to be skipped if pre_{op} fails? Should we still call $post_{op}$? This problem can be solved by providing a generalization that can be customized by plugins.

Pluggable Rule Engine Architecture

The pluggable rule engine architecture generalizes the current design and is fully backward compatible. The design provides a global policy enforcement that can be further customized for different semantics.

An example is that you can have error handling semantics encapsulated in a plugin, and by installing that plugin, you enable those error handling semantics. This requires the plugin architecture to load multiple rule engine plugins at the same time, and in a way that one plugin may provide semantics for another plugin.

Given a set of plugin operations, the pluggable rule engine architecture generates a PEP-added action Act as follows:

$$f : Op \rightarrow Act$$

$$f[op(args, env)] = pep(op, args, env)$$

To recover the default behavior, we can define pep as

$$pep : Op \times Args \times Env \rightarrow Act$$

$$pep[op, args, env] = pre_{op}(args, env); op(args, env); post_{op}(args, env)$$

We can define different error handling semantics as follows:

Skip $post_{op}$ if pre_{op} fails:

$$pep_1 : Op \times Args \times Env \rightarrow Act$$

$$pep[op, args, env] = if(pre_{op}(args, env) >= 0)\{op(args, env); post_{op}(args, env)\}$$

Run $post_{op}$ if pre_{op} fails:

$$pep_2 : Op \times Args \times Env \rightarrow Act$$

$$pep[op, args, env] = if(pre_{op}(args, env) >= 0)\{op(args, env)\}; post_{op}(args, env)$$

This way the rule engines form a hierarchy, with rule engines gradually refining the semantics of plugin operations. We can define such a hierarchy so that it is fully compatible with the current semantics, with the current rule engine at the bottom of the hierarchy, so that all existing rules run as expected. We can also, when new use cases arise, define a different set of plugins that implement different semantics, without changing the core code. This gives our users the flexibility to implement their policies.

Another challenge is the inter-rule-engine-call (IREC). Each rule engine provides a set of rules that it defines. Rules defined in one rule engine should be able to call rules defined in another rule engine. This is done through a universal callback function. The universal callback function is the only point of entry from the rule engine plugin to the iRODS core system. It handles all operations including accessing state information, accessing session variables, and the IREC. The general format of a callback is

$$fn(args)$$

where fn is a callback name and $args$ is a list of arguments. In the case of IREC, fn is the name of the rule and $args$ are the arguments to the rule. Compared to exposing a server API to the rule engine plugin, this approach has several advantages: First, this enables calling functions written in other programming languages as if they are microservices. Second, it allows us to add new APIs without changing the rule engine plugin interface. Third, we can add a pair of PEPs to this operation, which is sufficient for monitoring all interactions from the rule engine back to the core.

IMPLEMENTATION

The rule engine plugin architecture allows loading of multiple types of rule engine plugins, and multiple instances of each type of rule engine plugin. All instances share the same plugin object, but with different contexts. This way we don't have to load a rule engine plugin multiple times.

The rule engine contains the following four operations given in C++:

```
template<typename T>
irods:error start(T&);
template<typename T>
irods::error stop(T&);
template<typename T>
irods::error rule_exists(std::string, T&, bool&);
template<typename T>
irods::error exec_rule(std::string, T&, std::list<boost::any>&, callback);
```

The **start** function is called when the rule engine plugin is started. This happens when an iRODS process starts. The **stop** function is called when the rule engine is stopped. This happens when an iRODS process stops. The parameter is an object that can be used to pass data to and from these functions as well as other functions in the plugin operation. It can be thought of as the context. In fact, the state information can only be stored in this object. When the rule engine plugin manager loads more than one instance of the same plugin, the only object that is newly created is this object.

The **rule_exists** function accepts a rule name, a context, and writes back whether the rule exists in this plugin.

The **exec_rule** function accepts a rule name, a context, a list of arguments, and a callback object. The list of arguments are boxed by boost::any, and stored in a std::list container. This allows us to load the function in a dynamically linked library. The callback object is a C++ Callable, with the following interface method:

```
template<typename ...As>
irods::error operator()(std::string, As&&...);
```

The first parameter is *fn*. The second, third, etc. parameters are *args*.

One may have noticed that the callback interface expects raw values whereas the **exec_rule** function expects a list of values boxed by boost::any. Why do we design them like this? Ideally we would like to always use raw values to maximize efficiency, but this would require templates. We can accept raw parameters for the callback interface because it is statically compiled. But to allow the **exec_rule** to be loaded from a dynamic library, we cannot use templates. Because C++ templates are expanded at compile time, we cannot put a template function in a dynamically linked library that is linked to the main program at runtime. Wouldn't this be inefficient if the rule engine plugin simply wants pass the list of incoming arguments to the callback? The answer is to use the **unpack** construct as follows:

```
irods::error exec_rule(std::string _rn, T& _re_ctx, std::list<boost::any>& _ps, callback _cb) {
    cb(rn2, irods::unpack(_ps));
}
```

The **unpack** constructor is implemented so that the time complexity is $O(1)$.

The default implementation comes with a default rule engine. The default rule engine only has the *pep* rule and provides an implementation of the generalized PEP. It provides extended namespace support for the translation to the default semantics. Formally speaking, it implements the following function, given a list of n namespaces $ns_i, i \in \{1, \ldots, n\}$ (configured in `server_config.json`)

$$
\begin{aligned}
pep \quad &: \quad Op \times Args \times Env \to Act \\
pep[op, args, env] \quad &= \quad ns_1 pre_{op}(args, env); \ldots ns_n pre_{op}(args, env); \\
&\quad\quad op(args, env); ns_n post_{op}(args, env); \ldots ns_1 post_{op}(args, env)
\end{aligned}
$$

Here, for simplicity, we omitted error handling semantics.

By default, we have only one namespace which is $ns_1 = ""$, which implements the default semantics. We can implement different semantics outlined in the previous section by changing this plugin. We can add more namespaces and keep the default semantics. For example, we can add in another namespace for auditing $ns_2 = "audit\_"$ or indexing $ns_3 = "index\_"$. The rules listen to the *audit* namespace. For example pre and post file read PEPs are provided as follows:

$$audit\_pep\_resource\_read\_pre$$

$$audit\_pep\_resource\_read\_post$$

Rule engine plugins can be written to listen to those namespaces and provide the specific functionalities in a modular fashion. When a set of specialized plugins are installed, we can switch on/off a feature by just changing which namespaces are available.

APPLICATIONS
Python Rule Engine

We have created a proof of concept Python rule engine plugin. It allows users to implement PEPs directly in Python. This provides an avenue for the rapid expansion in the functionality of iRODS deployments, by taking advantage of the vast ecosystem of existing Python libraries as well as the large community of Python developers.

The plugin translates calls to `exec_rule` into calls to Python functions, whose implementations are loaded from `/etc/irods/core.py`, a Python code file. Users of the plugin are only required to write Python code, and are able to use all features of the Python programming language, including importing arbitrary Python modules.

Because of the pluggable rule engine architecture, this means iRODS users will be able to implement all PEPs directly in Python, or to call out to Python from other rule engine plugins, e.g. to extend the functionality of existing iRODS rules.

Event Tracking

The `audit` plugin provides an asynchronous tracking mechanism for every operation and their arguments and environments in iRODS, thereby providing a complete log. It runs at native code speed. Because the PEPs are dynamically generated, it supports any future plugin operation automatically. It allows the log to be sent to a remote system and processed asynchronously[1].

The rules listen to the *audit* namespace. To illustrate the implementation, a pre and post file read rule can be provided as follows:

---

[1] currently under development

```
audit_pep_resource_read_pre (...) {
    writeLine("serverLog", ...);
}
audit_pep_resource_read_post (...) {
    writeLine("serverLog",...);
}
```

In our implementation, these rules are implemented directly in C++ and therefore incur minimum overhead over normal operations.

CONCLUSION

We described a new development in the next release of iRODS. The pluggable rule engine architecture allows us to easily create new rule engines as plugins and run multiple rule engines concurrently. The pluggable rule engine architecture allows easy implementation and maintenance of rule engine plugin code and offers significant performance gains in some use cases. The pluggable rule engine architecture enables modular incorporation of features from other programming languages, allows efficient auditing of interactions of user-defined rules and the iRODS system, and supports full interoperability between rules and libraries written in different languages. This design allows us to easily incorporate libraries designed for different programming languages, for example, Python, C++, etc., into the policy sets, significantly enhancing the capabilities of iRODS without syntactic overhead. This new architecture enables a wide range of important applications including auditing, indexing, and modular distribution of policies.

QRODS: A Qt library for iRODS data system access

B. Silva, A. Lobo Jr., D. Oliveira, F. Silva, G. Callou, V. Alves, P. Maciel
Center for Informatics
UFPE, Recife, Brazil
{bs,aflj,dmo4,faps,grac, valn,prmm}@cin.ufpe.br

Stephen Worth
EMC Corporation
Massachusetts, U.S.A.
stephen.worth@emc.com

Jason Coposky
iRODS Consortium
Chapel Hill, U.S.A.
jasonc@renci.org

ABSTRACT

The evolution of the data center and data has been dramatic in the last few years with the advent of cloud computing and the massive increase of data due to the Internet of Everything. The Integrated Rule-Oriented Data System (iRODS) helps in this changing world with virtualizing data storage resources regardless the location where the data is stored. This paper explains and demonstrates a library that extends the Qt abstract model interface to provide access to the iRODS data system from within the Qt framework. Qt is widely used for developing graphical user interface software applications that are display platform agnostic. This library intends to benefit Qt developers by enabling a transparent iRODS access. Moreover, it will allow developers to implement applications that access an iRODS data system to populate a single model that can be displayed as a standard Qt tree like structure.

Keywords

iRODS, storage, RODS library, Qt, RODEX.

1. INTRODUCTION

The data center has evolved dramatically in recent years due to the advent of the cloud computing paradigm, social network services, and e-commerce. This evolution has massively increased the amount of data to be managed in data centers. In this context, the Integrated Rule-Oriented Data System (iRODS) has been adopted for supporting data management. The IRODS environment is able to virtualize data storage resources regardless of the location where the data is stored as well as the kind of device the information is stored on.

IRODS is an open source platform for managing, sharing and integrating data. It has been widely adopted by organizations around the world. iRODS is released and maintained through the iRODS Consortium [1] which involves universities, research agencies, government, and commercial organizations. It aims to drive the continued development of iRODS platform, as well as support the fundraising, development, and expasion of the iRODS user community. iRODS is supported by CentOS, Debian and OpenSuse operating systems. Since iRODS is an open source platform, the developed library must also support other Linux distributions.

For using iRODS, a few basic client tools are available such as: iCommands and iRODS Explorer. However, developers that would like to implement C++ applications that communicate with iRODS do not have any framework to conduct easy communication nor support for the development of graphical user interfaces. Therefore, it is not easy for developers to integrate iRODS with other software. This paper proposes a library to reduce this gap by providing support for Qt framework developers to communicate with iRODS.

In the last few years, some research has been performed to allow the communication of different storage systems or file systems. iRODS adopted a proprietary protocol to conduct the communication between storage nodes and

iRODS UGM 2015 June 10-11, 2015, Chapel Hill, NC

clients. Therefore, the development of customized client tool or the adoption of general tools to access iRODS data is difficult. The authors in [7] proposed other data transfer protocols such as WebSAV that is an open standard based on HTTP protocol. A generic interface, named Davis, was implemented to consider that open standard protocol. Davis is a WebDAV gateway to iRODS server regardless of its location. Besides that, experiments were conducted to show that the proposed approach did not impact the iRODS performance.

The main goal of [8] is to enable grid systems to access data from any arbitrary data source and to be able to transfer data between data sources with different protocols without the need of intermediate space such as a local storage space. In order to accomplish that, the authors designed a generic file system access framework as a backend to the GridFTP interface. The applicability of the proposed approach was demonstrated through a prototype named Griffin that was developed. That prototype considers iRODS data grid system as an example of an arbitrary data source.

In [6], the authors propose a framework that integrates data grid engines and network to facilitate complex policy-driven data operations. In order to test the proposed approach, the authors had combined OPenFlow rules with iRODS rules to allow non network expert users to easily access and control the network using the iRODS interface.

Different from the previous papers, the main goal of this work is to provide a library that allows Qt developers to implement a model that access the iRODS data system. Therefore, iRODS developers may adopt the proposed library for developing graphical user interface software applications with Qt. In order to accomplish that, we developed a mapping between iRODS API commands and Qt equivalents.

The proposed library extends QAbstractItemModel class and provides a model already integrated with the Qt model/view controller, Qt MVC, that access the iRODS data system. Qt is a framework widely used for developing software applications with graphical user interface [2]. The adoption of the proposed library allows Qt developers to implement applications that access iRODS data system through one sigle model that can be set to stardard Qt views, such as Qt tree view, Qt list view and Qt table view.

This paper is organized as follows. Section 2 briefly present the basic concepts needed for a better understanding about this work. Section 3 presents the developed library named QRODS. Section 4 describes an example implemented that uses QRODS library. Section 5 concludes the paper and makes suggestions on future directions.

2. PRELIMINARIES

This section presents important concepts for a better understanding of QRODS library. First, a brief overview related to iRODS is presented. Next, some concepts regarding QAbstractItemModel class and Jargon API are discussed.

2.1 iRODS

iRODS has become a powerful, widely deployed system for managing significant amount of data that requires extendable metadata. Typical file systems provide only limited functionality for organizing data and a few (or none) for adding to the metadata associated with the files retained. Additionally, file systems are unable to relate or structure what limited metadata is available and provide only a platform from which to serve unstructured file data. Within several fields, scientific research evolving instrumentation capabilities have vastly expanded the amount and density of unstrucuted file data, in which standard file systems can be a limiting factor in the overall use of data.

iRODS can be classified as a data grid middleware for data discovery, workflow automation, secure collaboration and data virtualization. As illustrated in Figure 1, the middleware provides a uniform interface to heterogeneous storage systems (POSIX and non-POSIX). iRODS lets system administrators roll out an extensible data grid without changing their infrastructure and accessing through familiar APIs. The reader should refer to [4] and [3] for more details about iRODS environment.

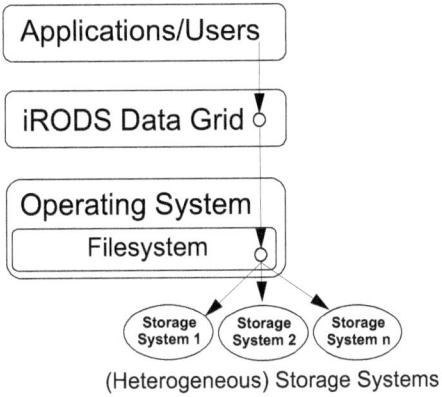

Figure 1. iRODS Overview [5].

2.2 QAbstractItemModel class

Qt is a comprehensive C++ application development framework for creating crossplatform GUI applications using a "write once, compile everywhere" approach. Qt lets programmers use a single source tree for applications that run on Windows, Linux, and Mac OS X as well as mobile devices. Qt libraries and tools are also part of Qt/Embedded Linux alike product that provides its own window system on top of embedded Linux.

The Qt framework provides a model/view controller approach (Qt MVC). QAbstractItemModel class provides an abstract interface for item model classes. Thus, programmers can populate one single model; such model enables the use of different ways for displaying a group of contents (files and directories). A QTreeView, for example, implements a tree representation of items from a model, whereas a QTableView implements a standard table representation.

2.3 Jargon API

Jargon is an API that implements the communication with iRODS protocol. The API allows development of iRODS-enabled Java applications. It is useful for developing mid-tier applications and services, as well as desktop-clients. These libraries also provide a foundation for a new set of interfaces that come with iRODS. Besides iRODS protocol implementation, Jargon is also able to acess iRODS data.

Jargon is implemented in Java, providing support for Java applications and not for other programming languages. Therefore, the iRODS Rest API based on Jargon has been developed to overcome such issue. The REST API provides support for developers to implement different client use cases for iRODS. Next session presents the QRODS, a library that adopts Jargon REST API to conduct the direct aceess from Qt applications to iRODS data system.

3. QRODS LIBRARY

This section presents the QRODS library. First, an overview of the library features is shown. Next, the architecture and the class diagram are presented.

3.1 Features

QRODS is a library that enables software engineers to build Qt graphical user interfaces (GUI) which can access the iRODS storage platform. The current version of QRODS implements essential manipulation functionalities of files and directories (collections). Therefore, the QRODS may perform the following proceedings:

- Create and delete files or collections;

- Download and upload files;

- Add and delete metadata from an object; and

- List content (files, metadata, collections).

3.2 Library Architecture

Figure 2 (a) depicts the QRDOS library architecture. As previously mentioned, the main goal of this library is to provide an interface that allow Qt developers to build applications with graphical user interfaces communicating to the iRODS data system. In order to accomplish that, Jargon REST API has been adopted to implement the iRODS protocol communication between our proposed library and iRODS. Jargon is an API that implements the communication with iRODS protocol. Although Jargon has been implemented in Java, it also provides a REST API for allowing tools implemented through different languages to adopt it to communicate with iRODS. Therefore, the QRODS library perform REST calls to the Jargon API that communicates with iRODS protocol through XML.

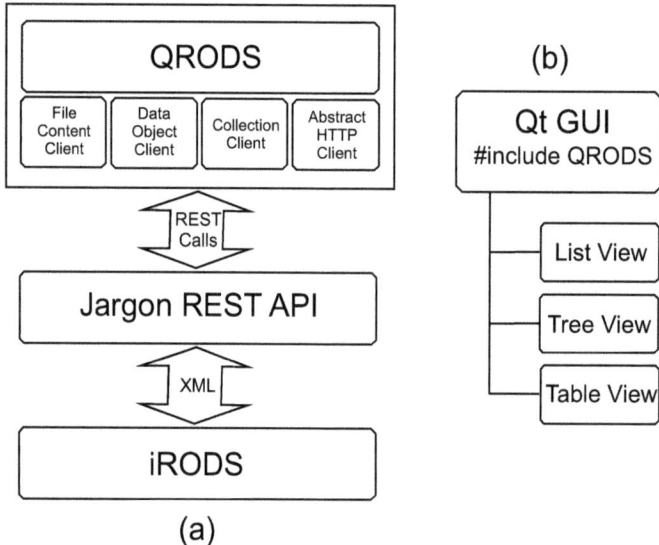

Figure 2. QRODS integrated with iRODS; (a) depicts the QRODS-IRODS access; and (b) highlights the three possibilities of iRODS content presentation.

The QRODS library has been composed of four different types of clients: (i) file content client, which is responsible to conduct all operations related to folders; (ii) data object client implements the metadata operations; (iii) collection client, which provides all functionalities for data collection; and (iv) abstract HTTP client, which provides secure communication functionalities. The following lines provide more details about the functionalities implemented for each type of client.

FileContentClient class

FileContentClient class manages iRODS files through two main functions: upload and download objects. As the name suggests, Qt applications upload files to iRODS using the `uploadFile()` function. The local and remote file names (including the entire file path) are parameters of that method. The equivalent iRODS command for this method is *iput*. Similar to uploading functionality, Qt applications download iRODS files by adopting the `downloadFile()` function. Local and remote paths represent the parameters used on that method. The equivalent iRODS command for this function is *iget*.

DataObjectClient class

DataObjectClient class manages metadata, which includes the add, remove and list functions. Qt applications associate a metadata to an iRODS object through the *DataObjectClient()* constructor. To remove an object metadata, the *removeDataObject()* method is adopted. A metadata can be listed by the *getDataObjectMetadata()* method or by its asynchronous version *getDataObjectMetadataAsync()*. The corresponding iRODS command for this method is *imeta*. In addition, a metadata collection can be added by the *addCollectionMetadata()* method.

CollectionClient class

CollectionClient class manages iRODS collections. This class contains methods to list collection contents, delete and create new collections. The equivalent iRODS commands to the CollectionClient class functionalities are *ils*, *irm* and *imkdir*. In order to delete a collection, the *removeCollection()* function must be called and the collection path is passed as parameter. Similarly, the *createCollection()* method receives the remote path as parameter to create new iRODS collections.

The listing functionality is implemented in two different ways:

- *Asynchronous Listing*: Using the *getCollectionDataAsync()* method, all the content of a collection is asynchronously listed. However, depending on the collection size, this function may take some time to finish.

- *Lazy Listing*: The asynchronous *getCollectionDataLazy()* method is adopted to perform collection lazy listing. By using this function, just a group of collection objects is retrieved per function call. This method is called several times to list all the collection objects. Therefore, this functionality is suitable for huge collections.

AbstractHTTPClient class

This class provides asynchronous functions associated with GET, POST, PUT AND DELETE HTTP calls. The `doGet()` method retrieves information identified by the requested URI. The `doPost()` method sends a post request with an enclosed entity to a given resource identified by the respective URI. The `doPut()` method requests the enclosed entity to be stored under the supplied requested URI. The `doDelete()` method requests that the resource identified by the requested URI to be deleted.

Qt Application Architecture using QRODS

Figure 2 (b) presents the architecture of a Qt application that includes the QRODS library to directly communicate with iRODS data system. Besides that, it is important to stress that the list, tree and table views were implemented in our QRODS library. Therefore, Qt developers may show the iRODS data as a list, a tree or a table view as depicted in Figure 3.

QRODS class Diagram

Figure 4 presents the QRODS class diagram. QAbstractItemModel is a Qt model view class which provides the abstract interface for item model classes. This class defines functions that are required to support table, trees and lists views. QAbstractItemModel class cannot be directly instantiated. Instead, a subclass must be implemented to create new models.

QRODS extends QAbstractItemModel overriding its main methods, such as: (i) *index()*, which returns the item model index specified by a given row, column and parent index; (ii) *parent()*, which returns the item model parent of a given index; and (iii) *headerData()* that returns the data for the given role and section in the header with the specified orientation.

QRODS is associated with one or more clients (e.g., Collection, FileContent or DataObject). The reader is redirect to Section 3.2 for more details about these clients.

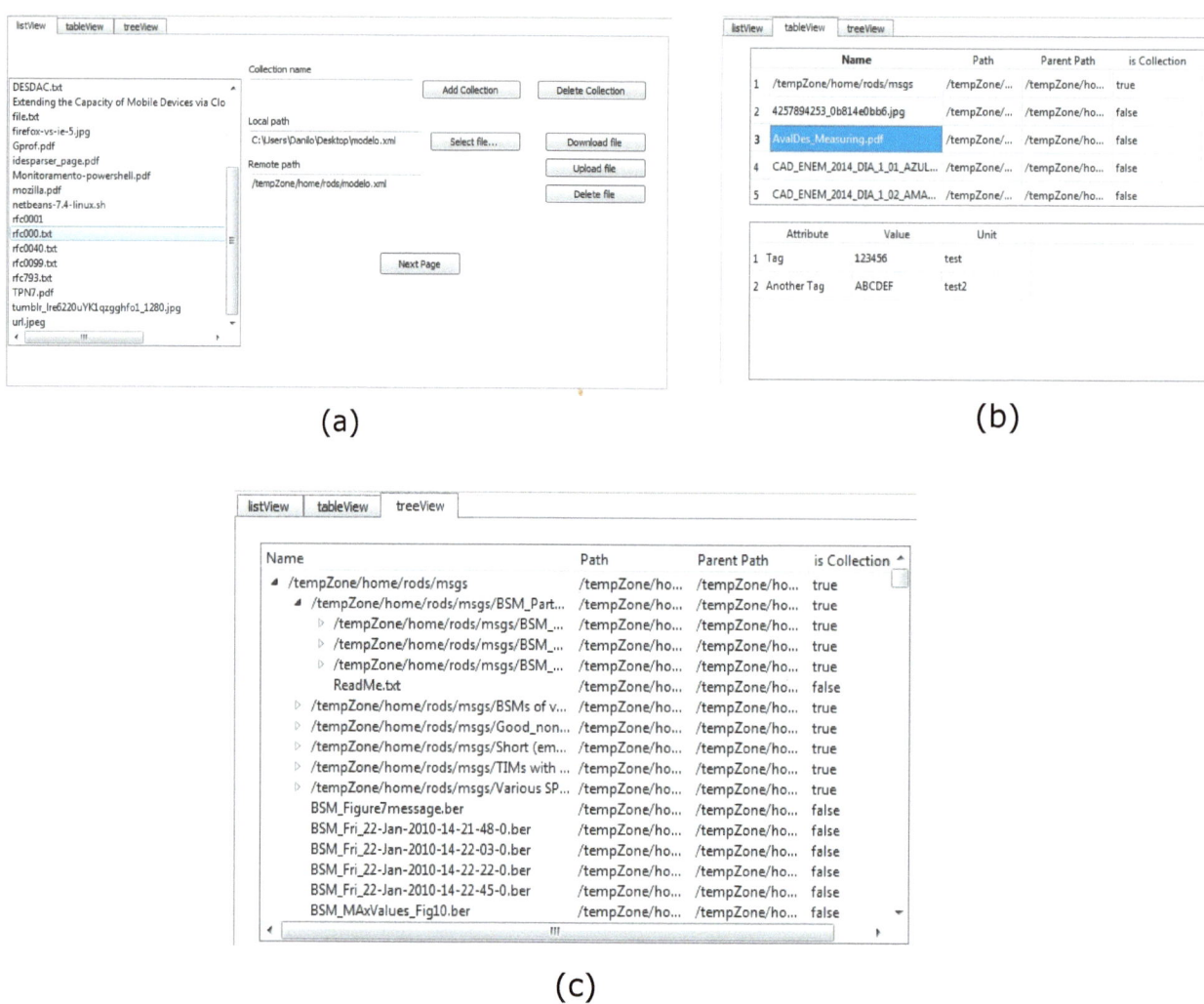

FileListingEntry represents one node in the file system (file or collection) having pointers to the father and child objects. FileListingEntry is used by the aforementioned three types of views (see Section 3.2).

AbstractHTTPClient provides secure communication through HTTP commands using SSL protocol and encrypted channels. In addition, passwords have been stored using AES 128 standard. MetadataEntry class encapsulates metadata information for a specific iRODS object. More specifically, it aggregates the corresponding attribute name, value and unit.

4. RODS EXPLORER - RODEX

The main goal of this section is to illustrate the applicability of the proposed QRODS library in an implemented Qt application. Therefore, RODs EXplorer (RODEX) was developed to show the main functionalities of our library. RODEX application is able to upload and download files, create and delete files or collections. This application also allow one to add, delete and list file metadata, and to list content (files, metadata or collections) of iRODS data system. In order to add such functionalities to the Qt framework, Qt developers just need to include QRODS library into their project.

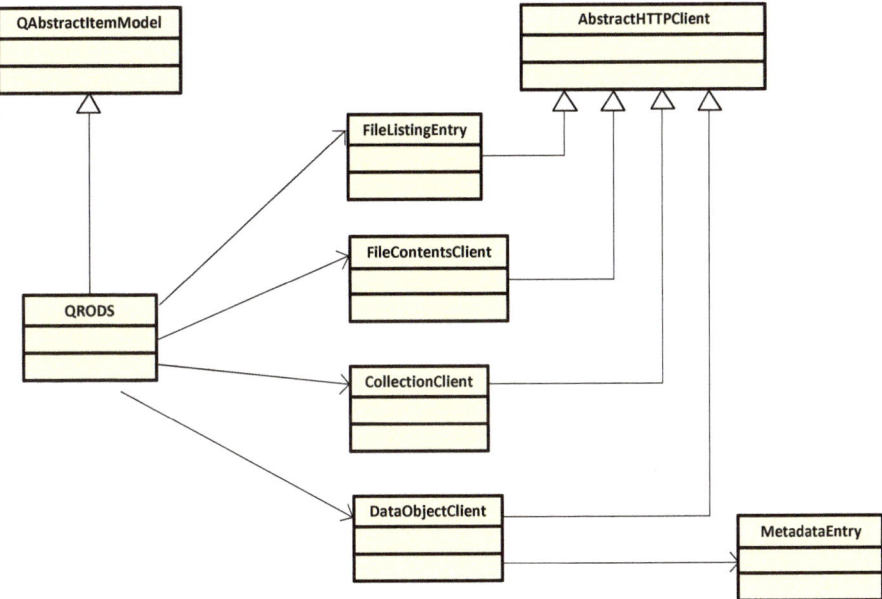

Figure 4. QRODS class diagram.

Figure 3 (a) depicts the RODEX application, a common Qt framework application using QRODS library that allows the direct access to iRODS collections and files. Users may add/delete a collection by providing the collection path. Related to files, it is possible to download, upload and delete files. For download and upload it is necessary to inform local and remote paths. Aiming to delete files only remote path field is required. Finally, button "Next Page" may be used when loading and visualizing collections at left side window. Files are loaded in low portions, obeying a fixed offset number.

Listing 1 shows two methods implemented in the MainWindow.cpp class for uploading and downloading files from iRODS data system. The main goal of this example is to show that the adoption of QRODS library is quite similar to the codes that Qt developers are used to implement. For instance, in the *on_downloadButton_clicked()* method (lines 2 to 10), it is just necessary to select the file to be downloaded from the list view as shown in Figure 3 (a). Besides that, the user must select the local path to download the file from the iRODS data system. The local path is achieved in line 4, and the remote path is represented through line 7. Afterwords, a call to the *getFile()* method is performed (line 9).

Lines 11 to 17 show the *on_uploadButton_clicked()* method that is responsible for uploading files to iRODS data system. For instance, to upload a file, a call to the *uploadFile()* method is performed (line 16). This method receives as parameters the local and remote paths. The local path is reached from the line 13 and the remote path from the line 14.

```
1  ...
2  void MainWindow::on_downloadButton_clicked()
3  {
4      QString localPath = ui->localPathEdt->text();
5
6      QModelIndexList list = ui->listView->selectionModel()->selectedIndexes();
7      QString remotePath = list.at(0).data().toString();
8
```

```
 9      fileClient ->getFile(remotePath, localPath);
10  }
11  void MainWindow::on_uploadButton_clicked()
12  {
13      QString localPath = ui->localPathEdt ->text();
14      QString remotePath = ui->remotePathEdt ->text();
15
16      fileClient ->uploadFile(remotePath, localPath);
17  }
18  ...
```

Listing 1. MainWindow.cpp example.

5. CONCLUSION

The evolution of data center due to the advent of cloud computing as well as the Internet of Everything has been increasing the amount of data to be managed by that systems. The iRODS helps in this changing world with virtualizing data storage resources regardless the location where the data is stored. This work has presented and demonstrated a library that extends Qt abstract model interface to provide access to the iRODS data system from within the Qt framework. This library, named QRODS, allows Qt developers to implement applications that access iRODS data system as a standard model and display it as a Qt tree like structure.

The developed library provides support for three different Qt views (TreeView, ListView and TableView) that allow Qt developers to manage iRODS data and metadata. Additionally, this library presents an asynchronous method to perform lazy collection listing for all supported views. Therefore, groups of collection objects are incrementally retrieved, which allows the use of QRODS to manage huge collections. Additionally, the RODEX application is implemented to show the applicability of the proposed QRODS library. As future directions, we intend to extend our QRODS library to deal with different remote access storage.

REFERENCES

[1] Irods consortioum web page, May 2015.

[2] Qt cross-platform web page, May 2015.

[3] EMC. Using an integrated rule-oriented data system (irods) with isilon scale out nas. http://www.emc.com/collateral/white-papers/h13232-wp-irodsandlifesciences-isilonscaleoutnas.pdf. Acessed: 04/25/2015.

[4] EMC. The integrated rule-oriented data system (irods). http://irods.org/, 2014. Acessed: 04/25/2015.

[5] EMC. irods executive overview. http://irods.org/wp-content/uploads/2014/09/iRODS-Executive-Overview-August-2014.pdf, 2014. Acessed: 04/25/2015.

[6] S. Huang, H. Xu, Y. Xin, L. Brieger, R. Moore, and A. Rajasekar. A framework for integration of rule-oriented data management policies with network policies. In *Research and Educational Experiment Workshop (GREE), 2014 Third GENI*, pages 71–72, March 2014.

[7] S. Zhang, P. Coddington, and A. Wendelborn. Davis: A generic interface for irods and srb. In *Grid Computing, 2009 10th IEEE/ACM International Conference on*, pages 74–80, Oct 2009.

[8] S. Zhang, P. Coddington, and A. Wendelborn. Connecting arbitrary data resources to the grid. In *Grid Computing (GRID), 2010 11th IEEE/ACM International Conference on*, pages 185–192, Oct 2010.

This paper originally appeared in The International Archives of the Photogrammetry, Remote Sensing and Spatial Information Sciences, Volume XL-1, 2014 ISPRS Technical Commission I Symposium, 17 – 20 November 2014, Denver, Colorado, USA.

RECOVER

AN AUTOMATED CLOUD-BASED DECISION SUPPORT SYSTEM FOR POST-FIRE REHABILITATION PLANNING

J. L. Schnase [a, *], M. L. Carroll [b], K. T. Weber [d], M. E. Brown [b], R. L. Gill [a], M. Wooten [b], J. May [d], K. Serr [d], E. Smith [d], R. Goldsby [d], K. Newtoff [c], K. Bradford [c], C. Doyle [c], E. Volker [c], and S. Weber [c]

[a] Office of Computational and Information Sciences and Technology,
[b] Biospheric Science Laboratory, and [c] DEVELOP Program Office
NASA Goddard Space Flight Center, Greenbelt, Maryland USA
(john.l.schnase, mark.carroll, molly.e.brown, roger.l.gill, margaret.wooten, kiersten.newtoff, kathryn.bradford, colin.s.doyle, emily.volker, samuel.j.weber)@nasa.gov

[d] Idaho State University GIS Training and Research Center, Pocatello, Idaho USA
(webekeit, mayjeff2, serrkind, smiteri6, goldrya2)@isu.edu

Pecora 19 / ISPRS TC 1 and IAG Commission 4 Symposium

KEY WORDS: Decision support systems, DSS, burned area emergency response, BAER, emergency stabilization and rehabilitation, ESR, cloud computing, rapid response

ABSTRACT:

RECOVER is a site-specific decision support system that automatically brings together in a single analysis environment the information necessary for post-fire rehabilitation decision-making. After a major wildfire, law requires that the federal land management agencies certify a comprehensive plan for public safety, burned area stabilization, resource protection, and site recovery. These burned area emergency response (BAER) plans are a crucial part of our national response to wildfire disasters and depend heavily on data acquired from a variety of sources. Final plans are due within 21 days of control of a major wildfire and become the guiding document for managing the activities and budgets for all subsequent remediation efforts. There are few instances in the federal government where plans of such wide-ranging scope and importance are assembled on such short notice and translated into action more quickly. RECOVER has been designed in close collaboration with our agency partners and directly addresses their high-priority decision-making requirements. In response to a fire detection event, RECOVER uses the rapid resource allocation capabilities of cloud computing to automatically collect Earth observational data, derived decision products, and historic biophysical data so that when the fire is contained, BAER teams will have a complete and ready-to-use RECOVER dataset and GIS analysis environment customized for the target wildfire. Initial studies suggest that RECOVER can transform this information-intensive process by reducing from days to a matter of minutes the time required to assemble and deliver crucial wildfire-related data.

1. INTRODUCTION

1.1 Background

Each year wildfires consume an average of 4.2 million acres of land in the United States, according to the National Interagency Fire Center. The long-term recent decade average is even higher (NIFC, 2013). Fire suppression activities have been employed since the early 1900s to preserve land and protect people and infrastructure. National coordination of fire suppression activities between federal agencies is performed by the National Interagency Fire Center in Boise, Idaho. Fire management begins when a fire is named and an incident command team is assigned; it progresses through the stages of fire suppression including initial and extended attack, followed by containment, control, and extinguishment. If necessary, when the fire is contained, burned area emergency response (BAER) teams may be assigned. These teams have seven days to make an assessment of post-fire conditions and develop a preliminary stabilization and rehabilitation; they have 21 days to submit a final plan once the fire is controlled (BAER, 2014).

Over the past two decades, major advancements have occurred in remote sensing technologies and geographic information systems (GIS). These Earth observational data and software have been employed by fire managers and those who support them to map and characterize fire locations and their extent. These maps can be combined with other geospatial data depicting resources, infrastructure and population centers to identify areas of strategic importance.

The majority of attention in mapping burned areas has historically focused on forested areas (Giglio, et al., 2009; Jirik, 2013; Kasischke, et al. 2011). However, the current project focuses on savanna fires, which research suggests can account for carbon emissions equivalent to or exceeding fossil fuel combustion by automobiles (Brustet, et al., 1992). For purposes of this study, savannas in the US are defined as semiarid grass and shrub dominated regions and are all located in the Western US (Figure 1). Much of these savannas are considered primary habitat for sage grouse, mule deer, and pronghorn antelope and are also used for livestock grazing. Fires can have profound short-term and long-term effects on the ecosystem.

---

* Corresponding author.

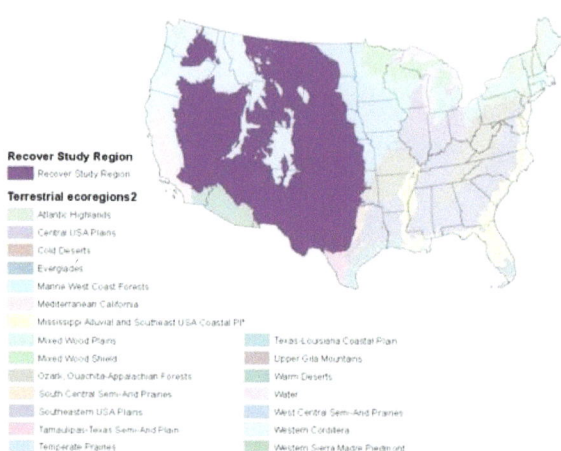

Figure 1. Savanna ecosystem to be served by the RECOVER system. *Image credit: Akiko Elders.*

The current process of preparing a fire rehabilitation plan typically begins with the BAER team leader requesting a fire severity product. A difference normalized burn ratio (dNBR) layer derived from Landsat imagery is generally the product that is delivered (Cocke, et al., 2005; Key and Benson, 1999). This product may be integrated with other data, such as topographic information, soil properties, land use, presence of threatened or endangered species, threats to life and property, historic and recent conditions of soil moisture, to create the knowledge base upon which a remediation plan is crafted.

This data assembly, analysis, and decision-making process must happen quickly in order to meet the statutory requirement of producing a preliminary BAER plan within seven days. However, right now, this process involves substantial human intervention, and the information gathered depends on the availability of staff, time, and data for a particular region. Even though there is a wide array of information services available to the wildfire community, these services tend to focus on research coordination, information sharing, fire risk assessment, active fire management, and fires on forested lands. None of the existing services address the specific needs of post-fire stabilization and restoration planning and monitoring vegetation recovery for semiarid lands.

To assist the effort to manage savanna fires, we are developing an automated decision support system (DSS) called the Rehabilitation Capability Convergence for Ecosystem Recovery (RECOVER) (Carroll, et al., 2013). This system compiles all the necessary datasets for the BAER teams rehabilitation planning and provides them in an easy to use web map interface. In this paper, we describe the RECOVER system and the RECOVER project, report on the results of our Phase 1 feasibility studies, and describe our future plans for operational deployment.

1.2 Challenge Being Addressed

The RECOVER project is focusing on the restoration of fire-impacted ecosystems as well as post-fire management and rehabilitation. The work is being funded by the NASA Applied Sciences Program and spans all four of the Program's primary themes of Health, Disasters, Ecosystem Forecasting, and Water Resources. Idaho State University's (ISU) GIS Training and

Research Center (GIS TReC, 2014) is the lead organization supported by the NASA Goddard Space Flight Center's Office of Computational and Information Sciences and Technology Office and Biospheric Sciences Laboratory. Our specific focus has been on the semiarid regions of the Western US (Sayre, et al., 2009), with RECOVER framed around the problems and challenges faced by the BAER program and the special requirements of post-wildfire decision-making with regard to reseeding in savanna ecosystems.

Wildfire is a common hazard throughout semiarid savanna ecosystems. Following fire, ground vegetation is typically eliminated, leaving the landscape devoid of cover. These communities may then experience a series of adverse changes due to landslides, soil erosion, and invasive plant infestations (Hilty, et al., 2004; Pierson, et al., 2002). While wildfires have occurred for millennia, the high frequency and intensity of today's wildfires contrast with those that occurred in the past (DeBano, et al., 1998; Thoren and Mattsson, 2002). These changes have led to unprecedented transformations to the semiarid savanna ecosystem.

Following wildfire, especially a high severity fire, the protective vegetation and organic litter cover are removed from hillsides, which can destabilize surface soils on steep slopes. Reseeding and other treatment approaches can rapidly stabilize the soil and promote water infiltration, thereby controlling erosion and preventing further loss of topsoil (Anderson and Brooks, 1975; Beyers, 2004; Miller, et al., 2003; Ribichaud, et al., 2006). Reseeding may also increase vegetation cover and forage availability for wildlife and livestock when appropriate initial plant establishments are used (Hubbard, 1975; Sheley, et al., 2997).

Given the importance of reseeding, it is not surprising that assessing the effects of wildfire, identifying areas that are likely to benefit from reseeding or other post-fire treatment, and monitoring ecosystem recovery in response to reseeding are important elements of BAER planning. However, as explained below, our initial feasibility evaluation has revealed that there also is significant interest in the use of RECOVER by active-fire incident response teams and by agency program managers for regional-scale, fire risk assessment.

1.3 Project Objectives

The primary objective of the RECOVER project has been to build a DSS for BAER teams that automatically brings together in a simple, easy-to-use GIS environment the key data and derived products required to identify priority areas for reseeding and monitor ecosystem recovery in savanna ecosystems following a wildfire. The fundamental propositions that have been tested during our Phase 1 feasibility study is whether RECOVER, the embodiment of such a system, substantively improves BAER team decision-making and, if so, whether the system can be deployed into practical use in the BAER program.[1] Beyond these fundamental questions, we have also used this Phase 1 feasibility study to identify unanticipated uses for the RECOVER technology and build the foundation for broader agency collaborations with the US Forest Service.

---

[1] Burned Area Emergency Response (BAER) is the name of the US Forest Service program; the corresponding program within the Bureau of Land Management is named Emergency Stabilitzation and Rehabilitation (ESR). For simplicity, we use BAER throughout in this paper.

1.4 Partner Organizations

Our primary partner organizations during the Phase 1 feasibility study have been the US Department of Interior (DOI) Bureau of Land Management (BLM) and the Idaho Department of Lands (IDL). BLM is the second largest agency in the eight-member National Interagency Fire Center the nation's wildfire coordinating center located in Boise, ID. BLM has operational responsibility for wildland fire on approximately 250 million acres of public land in the US, including approximately 12 million acres, or 22%, of the land base in Idaho. Idaho Department of Lands is the primary state-level agency responsible for dealing with wildfire in Idaho.

Since the RECOVER Phase 1 effort has focused on developing and evaluating capabilities in Idaho, the BLM/IDL teaming arrangement has been ideal for this feasibility study. In addition, partnering with BLM positions the project for broader regional- and national-scale operational deployment of RECOVER capabilities during Phase 2, as explained below. Likewise, interactions with NIFC through the BLM partnership open the possibility of RECOVER being adopted by other federal wildfire agencies.

ISU's GIS TReC has over many years developed a close working relationship with BLM and IDL in Idaho, which created a congenial and highly productive environment for this work. There has been significant involvement of key BLM and IDL collaborators at all stages of the project. We are working directly with, BLM's National Program Leader for Post Fire Recovery, as well as regional and state coordinators, field office personnel, and incident team leaders. Approximately one dozen individuals from the partner agencies have contributed to the Phase 1 study. Interactions with our agency collaborators by email and phone calls have taken place on a near daily basis. In addition, the project hosted a summer science team meeting and webinar that included demonstrations, training, and a field trip to Idaho fire sites.

2. THE RECOVER CONCEPT

2.1 Technical Approach and Innovations

The RECOVER DSS is made up of a RECOVER Server and a RECOVER Client (Figure 2). The RECOVER Server is a specialized Integrated Rule-Oriented Data System (iRODS) data grid server deployed in the Amazon Elastic Compute Cloud (EC2). The RECOVER Client is a full-featured Adobe Flex Web Map GIS analysis environment. When provided a wildfire name and geospatial extent, the RECOVER Server aggregates site-specific data from pre-designated, geographically distributed data archives. It then does the necessary transformations and re-projections required for the data to be used by the RECOVER Client. It exposes the tailored collection of site-specific data to the RECOVER Client through web services residing on the Server. This automatic aggregation can take place in a matter of minutes.

In a typical scenario-of-use, RECOVER uses the rapid resource allocation capabilities of cloud computing to automatically gather its various Earth observational and ancillary data products. Additional data can be added manually if needed, and the entire data collection is refreshed throughout the burn so that when the fire is contained, BAER teams have at hand a complete and ready-to-use RECOVER dataset that is customized for the target wildfire. The RECOVER server continues to gather data after the fire to support long-term monitoring of ecosystem recovery.

RECOVER's technical innovations are its use of cloud computing, data grid technology, and web services. Cloud computing provides agility and cost-savings, because RECOVER's Amazon cloud servers are an "elastic" resource that can be dynamically created and removed as needed. Another benefit to cloud computing is that RECOVER's compute and storage resources are acquired as an operational cost to the project, rather than through a time-consuming and potentially complex IT procurement: we simply pay Amazon for their services.

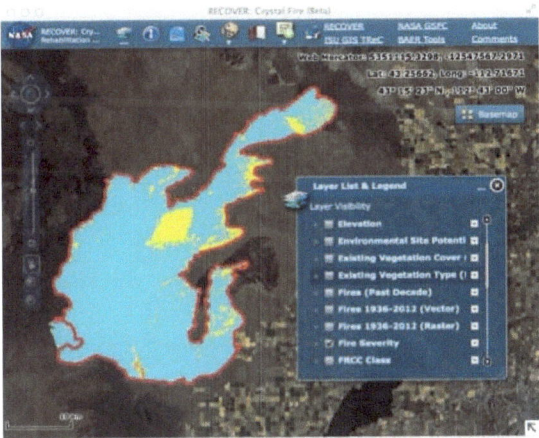

Figure 1. RECOVER Server and Client interfaces. For YouTube demonstrations, please see:
http://www.youtube.com/watch?v=LQKi3Ac7yNU   RECOVER Server
http://www.youtube.com/watch?v=SGhPpiSYpVE   RECOVER Client

Fire	Start Date	End Date	Acres Burned	RECOVER Response Time (min)	RECOVER Client URL
Crystal	15-Aug-06	31-Aug-06	220,000	N/A	http://naip.giscenter.isu.edu/recover/CrystalFire
Charlotte	2-Jul-12	10-Jul-12	1,029	N/A	http://naip.giscenter.isu.edu/recover/CharlotteFire
2 ½ Mile	2-Jul-13	3-Jul-13	924	30	http://naip.giscenter.isu.edu/recover/2nHalfMileFire
Mabey	8-Aug-13	19-Aug-13	1,142	120	http://naip.giscenter.isu.edu/recover/MabeyFire
Pony	11-Aug-13	27-Aug-13	148,170	35	http://naip.giscenter.isu.edu/recover/PonyFire
State Line	12-Aug-13	18-Aug-13	30,206	40	http://naip.giscenter.isu.edu/recover/StateFire
Incendiary Creek	18-Aug-13		1,100	90	http://naip.giscenter.isu.edu/recover/IncendiaryFire

Table 1. RECOVER feasibility study fires

The iRODS data grid technology at the core of the RECOVER Server enables the extensive use of metadata to manage files and individualize data collections to specific wildfire sites. It also provides a full-featured database capability for long-term archiving of all relevant information associated with a wildfire.

Our extensive use of web services allows RECOVER's site-specific data to be consumed by state-of-the-art web-based GIS applications, such as RECOVER's Adobe Flex Web Map Client. This makes it possible for our agency partners to avail themselves of RECOVER's analytic capabilities on any computer running a web browser, without having to acquire and maintain standalone GIS software. In addition, RECOVER's web services architecture facilitates the future development of client applications that run on mobile devices. Most modern smartphones, tablets, etc. actually consist of just the display and user interface components of sophisticated applications that run in cloud data centers. This is the mode of work that RECOVER is intended to eventually accommodate.

These innovations and RECOVER's overall architecture have reduced development costs, enabled tailored, wildfire-specific services, and reduced the amount of time required in the development process. It is important to note that these innovations will have similar effects on operational deployment going forward.

2.2 Application of Earth Observations

Earth observations and ancillary data play a crucial role in BAER decision processes. Key NASA observational inputs for this feasibility study have included Landsat 8, MODIS, and AMSR-E. These data are used to develop fire intensity, fire severity, NDVI, fPAR, ET, and many other products of specific use to the wildfire community (Weber, et al., 2008a,b). An important goal of the project, however, is to position the RECOVER system and the BAER community to be consumers of SMAP and Suomi NPP data.

RECOVER automatically gathers approximately two dozen core data products, including information on the fire site's vegetation cover and type, agroclimatic zone, environmental site potential, fire regime condition class, geology, hydrology, soils, historic fires, topography, and evapotranspiration. RECOVER also automatically assembles about two dozen historic, biophysical parameters that can be important in understanding pre-existing conditions. These include previous years' monthly averages for soil moisture, NDVI, temperature, precipitation, relative humidity, and surface winds. To support long-term monitoring, RECOVER automatically updates NDVI and fPAR data on a monthly basis post-fire.

Some data products require manual preparation. These are added to the RECOVER Server shortly after the automatic aggregation occurs and includes information about fire history, fire intensity, and habitats of importance to threatened or endangered species, and other types of idiosyncratic data relevant to a particular fire location. As described below, these enhancements reduce the time required for data assembly from days to a matter of minutes.

2.3 Application of Climate Model Outputs

The RECOVER project is breaking new ground by introducing reanalysis data into wildfire decision processes. RECOVER is acquiring its historic climatology data from the Modern Era Retrospective-Analysis for Research and Applications (MERRA) collection. MERRA uses NASA's GEOS-5 assimilation system to produce a broad range of climate variables spanning the entire satellite era, 1979 to the present, essentially integrating *the entire NASA EOS suite of observations* into key variables used by the RECOVER DSS (Reinecker, et al., 2011).

3. FEASIBILITY STUDY

3.1 Approach to Feasibility Assessment

During our Phase 1 feasibility study, we developed system requirements from a detailed study of the 2006 Crystal Fire, which burned 250,000 acres, making it one of the largest wildfires in Idaho's history. We used interviews, demonstrations, and reviewed decision-making processes and the resulting rehabilitation plans for the Crystal Fire with individuals who actually worked on the Crystal Fire BAER team. We then used an agile software engineering approach to build the RECOVER system, emphasizing at each step of the development process close customer involvement and rapid, incremental improvements to a continuously available system.

The ISU team was primarily responsible for providing GIS expertise, developing the RECOVER Client, building the test collection of Idaho datasets and web services, and working directly with the agency partners to provide training and respond to requests for specialized data products. The NASA team had primary responsibility for building the RECOVER Server, cloud computing, and providing expertise on Earth observational and MERRA reanalysis data.

The RECOVER system was deployed into experimental use on five active fires from the 2013 season: the 2 1/2 Mile, Mabey, Pony, State Line, and Incendiary Creek Fires (Table 1). Using RECOVER, we directly supported the work activities and data practices of agency colleagues as they responded to these fires, observing and gathering input from their experiences. Various ease-of-use, performance, and process factors were assessed.

BAER planning is a complex, multidimensional decision process. In order to accurately gauge the effects of our interventions, we assessed RECOVER's scientific, technological, and programmatic feasibility in terms of the following six criteria: *Science Feasibility* – (1) Does the RECOVER DSS work as well or better than current tools and methods in supporting the science and management decision making of BAER teams? (2) Is value added to BAER team decision processes by automatically assembling data and maintaining wildfire site-specific data in a central RECOVER Server?; *Technology Feasibility* – (3) Has RECOVER been integrated and tested in a relevant operational decision-making environment? (4) Is RECOVER able to produce required outcomes as fast or faster than conventional means?; and *Programmatic Feasibility* – (5) Can RECOVER reduce the cost or improve the effectiveness of data assembly, decision-making, and post-fire recovery monitoring? and (6) Is BLM and IDL willing to collaborate on the production development and operational deployment of RECOVER?

Criteria (1) and (2) were assessed qualitatively through daily interactions with individual BLM and IDL collaborators and as a group at the project's summer science team meeting. Successful prototype deployment was taken as the metric of success for criterion (3). Various ease-of-use, performance, and process factors were studied in the real-time context of the five active fires to assess criterion (4). Labor cost analyses for the historic and active fires were used to assess criterion (5). The level of interest and support of our agency collaborators in operationalizing RECOVER was the metric used to assess criterion (6).

3.2 Results of the Feasibility Study

RECOVER's *science feasibility* was strongly substantiated by the feedback received on criteria (1) and (2). More than one dozen agency collaborators provided input into this study, including individuals directly responsible for gathering and analyzing wildfire-related data, such as GIS analysts, incident response team members, and natural resource managers, as well as senior managers responsible for long-term national-scale programmatic development.

The overwhelming response to RECOVER has been positive. The consensus view is that there is great value in RECOVER's ability to automatically collect data that would otherwise be assembled by hand. In addition, having a single data management environment where all relevant information about a fire can be maintained over time and easily accessed significantly improves standard practice. The RECOVER Client Web Map application is easily accessed on the web from any workstation and provides a comprehensive feature set for routine GIS analytics without the need to maintain stand-alone GIS applications on the workstation. This is a significant convenience and has the potential for substantially reducing the cost and complexity of IT systems administration for our partner agencies.

RECOVER's *technical feasibility* has been demonstrated by affirmative results for criteria (3) and (4). We have successfully validated the system in the context of active fires, which has allowed us to effectively estimate potential improvements to the target decision-making processes and project impacts on cost,

functionality, and delivery options. The most dramatic example of RECOVER's capacity for process improvement is in the significant reduction in the time required to gather wildfire data into a GIS analysis environment. Where it used to require a day or two of work and special knowledge about the data needed and where to retrieve the data, RECOVER's automatic aggregation coupled with minimal manual additions of data has reduced that time to minutes or hours (Table 1) and lowered risk by reducing the dependence on institutional knowledge.

RECOVER's *programmatic feasibility* has been demonstrated by affirmative results for criteria (4) and (5). Over the past four years, there have been on average 120 wildfires per year in Idaho and over 200 in 2012 alone. Looking at the data assembly task alone for these fires, and assuming that at least one GIS analyst for each partner agency takes a day or two to collect data and perform initial assessments, labor impacts could run as high as 4000 hours or more, nearly 2.0 FTEs. While this is substantial, we believe it represents only a fraction of the time that could be saved on data assembly and data management over the full information lifecycle of fire-related data.

The most significant programmatic impacts, however, are likely to be on the improved quality of science understanding and management decision-making that can result from shifting valuable staff resources away from the mundane task of data gathering to the crucial jobs of analysis, planning, and monitoring. These benefits are almost impossible to quantify. The process enhancements realized by RECOVER offer the prospect of fundamentally improving the quality of our partner's work practices. For this reason, there has emerged at all levels within our partner agencies strong support for the RECOVER project and a desire to move forward with operational deployment of these capabilities. We therefore satisfy criterion (6), which has led to a successful NASA Applied Sciences Program review and the decision to fund Phase 2 implementation of the RECOVER system.

4. OPERATIONAL DEPLOYMENT

Over the next three years, we will deploy RECOVER into operational use. Starting with the Great Basin states of Idaho, Utah, and Nevada, we will ultimately support all of the Western US. Initially, our primary customer will continue to be BLM and the state-level agencies responsible for wildfire response, stabilization, and rehabilitation, but the goal is to grow the partnership to include the US Forest Service, National Park Service, and other state and federal agencies that have a need for these capabilities. The Great Basin focus is at the specific request of BLM and represents a scale of development that we believe assures technical, economic, and political success in the early going. We will initially continue with our focus on savanna ecosystems, but the goal will be to extend RECOVER's capabilities to support forest ecosystems as well.

4.1 Baseline Conditions

Currently, regional- and state-scale *fire risk assessment* is a largely ad hoc process carried out by BLM program managers. It is a type of decision process that is crucial to the agency and influences long-range resource planning. However, the process now involves a great deal of manual data assembly, the gathering of information from a diverse and often difficult-to-access suite of online wildfire information resources, and integrating this information in useful ways into stand-alone desktop GIS applications.

This paper originally appeared in The International Archives of the Photogrammetry, Remote Sensing and Spatial Information Sciences, Volume XL-1, 2014 ISPRS Technical Commission I Symposium, 17 – 20 November 2014, Denver, Colorado, USA.
Reprinted with permission of the authors.

Emergency preparations for *active-fire response* by incident managers involves the rapid gathering of data about the incident site prior to on-site arrival. Currently, the technical response can vary by incident team based on the specific training and experience of team members. Most incident teams have access to basic terrain maps, maps of roads and streams, etc., but use of advanced remote sensing and GIS capabilities is still not uniform between teams. With RECOVER, we will be able to arm response teams in advance with a complete collection of historic and current environmental data without depending on the incident team having a high-level GIS technician. By automatically generating data layers that are useful in understanding fire intensity and patterns of potential spread, we enhance the ability of incident teams to respond to the fire.

Post-fire rehabilitation plans of the sort developed by ESR and BAER teams are generally developed by multi-agency teams of specialists that include natural resource managers and scientists with expertise in various disciplines. Remote sensing imagery is used to complement field-based assessments, and several indicators derived from satellite imagery are used to characterize fire severity and intensity. For the current baseline condition, data collection in this setting is also a largely manual process performed by GIS analysts and resource managers. There is often limited access to critical data at the wildfire site, since again, most analysis environments are stand-alone applications running on field office workstations.

During *post-fire recovery monitoring*, fire managers are responsible for measuring and reporting the results of treatment plans. In the current baseline condition, this is generally carried out by field sampling. As a result, detailed, comprehensive evaluations are limited by budget and staff resources. PI Weber has demonstrated, through NASA-funded research that large-scale post-fire vegetation productivity in response to re-seeding can be monitored effectively using time-series NDVI and fPAR measurements (Chen, et al., 2012). RECOVER's ability to automatically assemble these post-fire metrics will significantly reduce the cost associated with recovery monitoring and improve the quality of our scientific understanding of what leads to successful treatments.

4.2 Implementation Approach and Key Milestones

We will focus on enabling four key work processes: fire risk, active-fire, and post-fire decision making and long-term recovery monitoring. Our approach will be to create a RECOVER deployment package — a collection of production-hardened technologies, technical documentation, training materials, and data services — that can become a model and mechanism whereby BLM and other agencies can replicate the deployment of RECOVER in other states.

Our implementation strategy will involve system development, data development, and operational deployment. The major work to be accomplished in *system development* is to move RECOVER V0.9, the experimental Client/Server platform used for Phase 1 feasibility analysis, to a fully functioning V1.0 of the system. We also will develop a mobile RECOVER Client that can run on iPads and iPhones (and Android clients if required by the customers). We will continue to use the agile software development approach that has proved successful in Phase 1, emphasizing close collaborations with our customers at all stages of development. Service level agreements (SLAs) will be defined, and the RECOVER Server and Clients will be security and performance hardened, validated, and tested as

needed in order to meet the requirements of a NASA/BLM-defined operational readiness review (ORR).

The *data development* work of Phase 2 will focus on completing RECOVER's core data catalog, which requires building a comprehensive base set of automatically retrievable data products, including the entire suite of MERRA variables and SMAP Level 4 soil moisture data when they become available. We also will work with our agency partners to develop specialized fire risk and post-fire data products of the sort described on in Section 4.4.

As described above, for *operational deployment*, we intend to design and develop a RECOVER deployment package that contains all the RECOVER components required to set up RECOVER on a state-by-state basis. The motivation for this is BLM's success in rolling out other new capabilities using this approach: the agency often partners with universities or other private-sector, non-profit entities at a state or regional level in order to accomplish its mission. BLM's long-standing partnership with ISU's GIS TReC in Idaho is an example of such a relationship. We will hold yearly science team meetings and define six-month milestones based on the major development threads described above.

4.3 Potential Impacts

According to our agency partners, data assembly is the most significant bottleneck in wildfire-related decision making. RECOVER dramatically reduces the time it takes to gather information, and it delivers the information in a convenient, full-featured, web-based GIS analysis environment. As a result, RECOVER's most important impacts are likely to be on the improved quality of science understanding and management decision-making that results from shifting valuable staff resources away from the task of data gathering to the more important tasks of analysis, planning, and monitoring. The potential magnitude of the resource shift is significant. If we extend to our three Great Basin states the labor impact example shown for Idaho in Section 3.2 above, as many as 6.0 FTEs per fire season could be mobilized in more productive ways.

4.4 Science Activities

RECOVER's ability to rapidly gather historic observational data and reanalysis climatology data along with measurements of current conditions makes it possible to do precision analyses that simultaneously inform our understanding of fire risk, active-fire fuel conditions, and post-fire rehabilitation potential. For example, for 2013's Pony Fire, we were able to use NDVI time series data to identify an early flush of vegetation in late February that fell off to a significantly lower-than-average level by June, suggesting there may be substantial senescent vegetation in this region: fine fuels that increase susceptibility to ignition. Also, 2013's mid-June NDVI levels were reduced by more than 10 percent compared with those over the previous decade, further confirming the anomaly (Figure 2). Being able to perform this type of analysis over automatically assembled data is seen as a potential game-changer by our partners.

RECOVER's potential as a research platform has led to a teaming arrangement with NASA's DEVELOP Program. DEVELOP fosters rapid feasibility projects that employ NASA Earth observations in new and inventive applications, demonstrating the widest array of practical uses for NASA's Earth observing suite of satellites and airborne missions.

This paper originally appeared in The International Archives of the Photogrammetry, Remote Sensing and Spatial Information Sciences, Volume XL-1, 2014 ISPRS Technical Commission I Symposium, 17 – 20 November 2014, Denver, Colorado, USA.

Reprinted with permission of the authors.

DEVELOP application projects serve the global community by cultivating partnerships with local, state, regional, federal, and academic organizations, enhancing the decision and policy making process and extending NASA Earth Science research and technology to benefit society.

As many as four DEVELOP scientists will be working with the RECOVER project. The DEVELOP team will conduct research on the identification and quantification of anomalous conditions that contribute to the risk of fire and that influence post-fire recovery using remotely sensed data and climate model outputs. The specific objective of the DEVELOP science activities is to extend the RECOVER DSS to ingest and produce products that incorporate SMAP, Suomi NPP, and other observational data, along with MERRA reanalysis data for improved decision making and extension of RECOVER to new end-users, such as the US Forest Service and the National Park Service.

4.5 Other Contributions of the Project

The RECOVER project is working with Esri's Landscape Analysis project in what promises to be a valuable off-shoot of these efforts. The Landscape Analysis (LA) program comprises a suite of data services designed to enable better informed decisions related to natural resources. We are working with Esri to use applicable LA layers in RECOVER, and, in turn, are contributing high-resolution RECOVER data and MERRA reanalysis data to Esri's program. This partnership highlights RECOVER's role as a broker for valuable wildfire-related data and extends RECOVER's impact to the broader GIS community.

5. CONCLUSIONS

The RECOVER DSS is an automated data assembly and web map decision support system that allows BAER teams to expedite the process of creating post-fire rehabilitation plans. Data are assembled for specific sites simply by providing a fire name and its spatial extent. Additional data sets can be added manually, and all data are maintained in a common format for ease of use. Initial prototype evaluations have demonstrated the effectiveness of RECOVER in reducing the time required to gather and deliver crucial data.

The RECOVER system is being extended to include research products from the fire science community and through the DEVELOP program. These extensions provides an opportunity to get new products to fire managers for evaluation and potential incorporation into management strategies. The agile architecture of RECOVER allows for incorporation of new datasets to the system by simply adding the web service to a list of "standard" products that are acquired in the data acquisition phase. This makes it possible to tailor RECOVER offerings to client needs.

The RECOVER decision support system will be operationally deployed over the next three years and available for use by federal and state agencies across the western United States. In this process, the RECOVER team is actively seeking input from the fire management community. Our goal is to work closely with end users to adapt the RECOVER system to the real and pressing needs of the fire response community.

ACKNOWLEDGEMENTS

This work represents the crucial and much appreciated contributions of many people. We thank our collaborators in the Bureau of Land Management and Idaho Department of Lands for their help designing and evaluating the RECOVER DSS. We are especially indebted to M. Kuyper, S. Jirik, and D. Repass for their support and leadership on the project. B. Holmes, A. Webb, B. Dyer, G. Guenther, T. Lenard, Z. Peterson, J. Nelson, D. Booker-Lair, R. Dunn, E DeYong, A. Mock, and A. Andrea provided helpful input throughout. We appreciate the long standing support and encouragement of R. Schwab at the National Park Service. D. Duffy, M. McInerney, D. Nadeau, S. Strong, and C. Fitzpatrick at NASA Goddard Space Flight Center provided crucial technical support, as did G. Haskett, T. Gardner, and K. Zajanc at ISU's GIS Training and Research Center. We thank L. Childs-Gleason, J. Favors, and colleagues in the NASA DEVELOP National Program Office for their help connecting the RECOVER project to the DEVELOP mission. Finally, we wish to thank L. Friedl, V. Ambrosia, and A. Soja in the NASA Applied Sciences Program for their support, programmatic leadership, and encouragement. Funding for this project has been provided by the NASA Science Mission Directorate's Earth Science Applied Sciences Program.

REFERENCES

Anderson, W. E., and L. E. Brooks. 1975. Reducing erosion hazard on a burned forest in Oregon by seeding. *Journal of Range Management* 28: 349-398.

Beyers, J. L. 2004. Postfire seeding for erosion control: effectiveness and impacts on native plant communities. *Conservation Biology* 18: 947- 956.

Brustet, M., Vickos, J. B., Fontan, J., Podaire, A., and Lavenu, F. 1992. Characterization of Active Fires in West African Savannas by Analysis of Satellite Data: Landsat Thematic Mapper. *In Global Biomass Burning*, J. S. Levine, Ed., Cambridge Massachussets: MIT Press, pp. 53-60.

Carroll, M. L., Schnase, J. L., Weber, K. T., Brown, M. E., Gill, R. L., Haskett, G. W., and Gardner, T. A. 2013. A New Application to Facilitate Post-fire Recovery and Rehabilitation in Savanna Ecosystems. *http://www.earthzine.org/2013/06/22/a-new-application-to-facilitate-post-fire-recovery-and-rehabilitation-in-savanna-ecosystems/*.

Chen, F., Weber, K.T., and Schnase, J.L. 2012. Assessing the success of post-fire reseeding in semiarid rangelands using Terra MODIS. *Rangeland Ecology and Management*, Vol. 65, No. 5, pp. 468-474.

Cocke, A. E., Fule, P. Z., and Crouse, J. E. 2005. Comparison of burn severity assessments using Differenced Normalized Burn Ratio and ground data. *International Journal of Wildland Fire*, 14: 189-198.

DeBano, L. F., D. G. Neary, and P. F. Pfolliott. 1998. *Fire's Effects on Ecosystems*: New York, John Wiley and Sons, 333 pp.

DEVELOP, 2014. *http://develop.larc.nasa.gov.*

Giglio, L. Loboda, T., Roy, D. P. Quayle, B. and Justice, C. O. 2009. An active-fire based burned area mapping algorithm for the MODIS sensor. *Remote Sensing of Environment,* 113: 408-420.

Hilty, J. H., D. J. Eldridge, R. Rosentreter, M. C. Wicklow-Howard, and M. Pellant. 2004. Recovery of biological soil crusts following wildfire in Idaho. *Journal of Range Management* 57: 89-96.

Hubbard, W. A. 1975. Increased range forage production by reseeding and the chemical control of knapweed. *Journal of Range Management* 28: 406-407.

Idaho State University GIS Training and Research Center (GIS TReC), *http://giscenter.isu.edu.*

Jirik, S. 2013. "Personal Communication," K. Weber.

Kasischke, E., Loboda, T., Giglio, L., French, N. H. F., Hoy, E. E., de Jong, B., and Riano, D. 2011. Quantifying burned area from fires in North American forests – implications for direct reduction of carbon stocks. *Journal of Geophysical Research,* 116.

Key. C. H., and Benson, N. C. 1999. The normalized burn ratio (NBR): A Landsat TM radiometric index of burn severity. *http://www.nrmsc.usgs.gov/research/ndbr.htm.*

Miller, J. D., W. J. Nyhan, and S. R. Yool. 2003. Modeling potential erosion due to the Cerro Grande Fire with a GIS-based implementation of the Revised Universal Soil Loss Equation. *International Journal of Wildland Fire* 12: 85-100.

National Interagency Fire Center (NIFC). 2014. *www.nifc.gov.*

Pierson, F.B., D.H. Carlson, and K.E. Spaeth. 2002. Impacts of wildfire on soil hydrological properties of steep sagebrush-steppe rangeland. *International Journal of Wildland Fire* 11:145-151.

Rienecker, M.M., & Coauthors. 2011. MERRA: NASA's Modern-Era Retrospective Analysis for Research and Applications. *Journal of Climate,* Vol. 24, No. 14, pp. 3624–3648. Available online at *http://dx.doi.org/10.1175/JCLI-D-11-00015.1.*

Robichaud, P. R., T. R. Lillybridge, and J. W. Wagenbrenner. 2006. Effects of post-fire seeding and fertilizing on hillslope erosion in north-central Washington, USA. *Catena* 67: 56-67.

Sayre, R., P; Comer, H. Warner, and J. Cress. 2009. *A new map of standardized terrestrial ecosystems of the conterminous United States,* US Geological Survey Professional Paper 1768, 17 p. (Also available online.)

Sheley, R. L., B. E. Olson, and L. L. Larson. 1997. Effect of weed seed rate and grass defoliation level on diffuse knapweed. *Journal of Range Management* 50: 39-43.

Thoren, F., and D. Mattsson. 2002. Historic wildfire research in southeastern Idaho. *http://giscenter.isu.edu/research/techpg/blm_fire/historic/wildfire_report.pdf.*

Weber, K. T., S. Seefeldt, and C. Moffet. 2009. Fire severity model accuracy using short-term, rapid assessment versus long-term, anniversary date assessment. *GIScience and Remote Sensing.* 46(1): 24-38

Weber, K. T., S. Seefeldt, C. Moffet, and J. Norton. 2008. Comparing fire severity models from post-fire and pre/post-fire differenced imagery, *GIS Science and Remote Sensing* 45(4): 392-405

Weber, K. T., S. S. Seefeldt, J. Norton, and C. Finley. 2008a. Fire severity modeling of sagebrush-steppe rangelands in southeastern Idaho. *GIScience and Remote Sensing* 45: 68-82.

iRODS Technology Applied to the DataNet Federation Consortium

Mike Conway
DICE-UNC
Chapel Hill, NC
michael_conway@unc.edu

Hao Xu
DICE-UNC
Chapel Hill, NC

Dr. Reagan Moore
DICE-UNC
Chapel Hill, NC

[1] ABSTRACT

The DataNet Federation Consortium (DFC) is one of the DataNet projects funded by the National Science Foundation to create national-scale cyber-infrastructure to support collaborative research. The DFC infrastructure "provides a new approach for implementing data management infrastructure that transcends technology, social networks, space, and time through federation-based sustainability models." [1] iRODS technology [2] is at the core of the federated, policy-managed data infrastructure of DFC, and DFC efforts are creating new capabilities to support shared research, and long term data management that are of interest to the broader iRODS community.

Keywords

DataNet, iRODS, Jargon

[2] INTRODUCTION

The DataNet Federation Consortium is developing national scale cyber-infrastructure to manage scientific data through its entire life cycle. At the heart of DFC is the concept of shared research collections that may be shared, manipulated through computational processes and transformations, discovered, and preserved for the long term. [3] Moore describes a life cycle for research collections that describe the policies and properties that distinguish one phase of the life cycle from another, as expressed in this diagram:

Project Collection	Data Grid	Data Processing Pipeline	Digital Library	Reference Collection	Federation
Private	Shared	Analyzed	Published	Preserved	Sustained
Local Policy	Distribution Policy	Service Policy	Description Policy	Representation Policy	Re-purposing Policy

Stages correspond to addition of new policies for a broader community
Virtualize the stages of the collection life cycle through policy evolution

Implied in this scheme are basic capabilities in multiple categories, including metadata management for discover-ability, integration of data processing and work-flows, and methods for federating and sharing data that have been discovered and assembled into collections. Central to this life-cycle concept is the idea that collections are defined by different types of management policies at each stage. The research group that assembles a project collection, typically shares tacit knowledge about the data context. As the data are shared with broader communities, the tacit knowledge is made explicit in the form of metadata. Policies are developed for each new community to enforce acquisition of descriptive and provenance metadata. These policies are enabled by capabilities being developed within the DFC infrastructure to enhance the discovery, sharing, re-purposing, and ultimate preservation of research data.

DFC is a cyber-infrastructure project. Gannon et al. have and appropriate definition that maps nicely to the above identified DFC capabilities, specifically, they identify five components. [4]

1. Data search and discovery tools.

2. Security.

3. User private data storage.

4. Tools for designing and conducting computational experiments.

5. Data provenance tracking.

These components correlate well to specific new capabilities being developed under DFC. However, we would like to extend this definition to a sixth component, namely, that cyber-infrastructure includes integration. More specifically, cyber-infrastructure exists as a middle-tier capability, but can hardly be effective as an island or walled garden. Data that cannot be accessed efficiently cannot be shared or discovered. Therefore, a key capability of cyber-infrastructure is the ability to integrate the above capabilities into a larger computational framework through API, standard protocols, as well as user interfaces of various types. A great deal of effort in DFC is in this 'sixth' component, however, this integration and interface concern will not be covered in this paper.

[3] DISCUSSION

Gannon's Components of Cyber-infrastructure as mapped to DFC and iRODS

If we use Gannon's definition of cyber-infrastructure as a template, we can explore various new technologies that are being built with the iRODS server at the core. We can use this as a basis for discussion, relate it to DFC, and highlight features of the DFC cyber-infrastructure approach.

Data Search and Discovery Tools

Metadata Templates

Data search and discovery tools are key to sharing and re-use of scientific data collections. The DFC grid, through the application of policies appropriate to the life-cycle stage, is concerned with preserving the significant properties of the data, both through the long-term preservation of curated metadata, and through the ability to automatically extract and maintain metadata based on the application of computational processes and procedures. Significantly, Moore's life-cycle concept identifies the act of creating new collections based on shared, federated data, and this ability to arrange new collections based on existing collections is essentially treated as a data search and discovery activity, as we will see in the developmetn of virtual collections.

An important, novel development in iRODS to support the human curation of metadata is the concept of metadata templates. Simply stated, metadata templates allow the definition of required and optional metadata values, including user friendly naming and documentation, validation and type information, as well as mapping information to store metadata inside the iRODS catalog. These metadata templates are defined and stored in some persistent store, and bound to collections in iRODS. Once associated with a collection, metadata templates allow various interesting capabilities.

First, metadata templates allow, through the definition of user cues and validation information, including default values, ranges, and options, the dynamic generation of user interface components. This can be for information re-display, allowing nicely formatted, user-friendly displays instead of simple lists of metadata attributes. Users may also be provided with data entry forms that include validation logic based on template contents.

Metadata templates are metadata themselves, allowing structure to be added to otherwise unstructured iRODS AVU metadata. AVUs in iRODS are simply attribute-value-unit triples attached to files, collections, resources, users, and other iRODS catalog objects. These unstructured metadata values can be modified through facilities such as the imeta iCommand [5] and rules that query and update AVUs.

Metadata templates provide grouping of metadata along with information on meaning of assigned AVU triples. metadata templates are identified by unique UUIDs, and each element is also named with a UUID. This device allows templates to include elements that are linked to a master catalog of elements. This can assist in reuse and crosswalks of data between metadata schemes.

Metadata template design contemplates a distinction between user curated and automatically generated metadata. Certain templates, instead of resulting in the generation of user entry forms, can indicate that a computational process will derive metadata values. In this scenario, the template holds a reference to the computational process that will do the extraction, and will provide a mapping in the template elements that can bind an output parameter of the extraction process to AVU values. Using this approach provides a declarative way of specifying metadata extraction policies that are dynamically applied based on a per-collection resolution mechanism.

Like many new DFC features, we are taking the attitude of providing an abstract definition of the service, married to a 'reference implementation'. This reference implementation is meant to utilize only iRODS facilities, meaning that metadata templates are stored as JSON structures decorated with AVUs, and the binding and discovery process depends on AVUs applied to collections and data objects. The abstract interfaces [6] and reference implementation [7] may be found in the DFC GitHub repository.

iRODS Indexing Framework

A major innovation in the iRODS architecture is the definition of an indexing framework as an extension of the iRODS policy-managed approach. This has already been demonstrated in a prototype form using the HIVE system to apply SKOS vocabulary terms to iRODS [8], and for full-text indexing using the ElasticSearch platform. The basic components include:

1. Extensions to the iRODS policy enforcement points [9] so that events may be captured as standard JSON representations. For example, storing a file in DFC triggers a PostProcForPut event, and that event can be represented with information describing the data that was stored on the grid. Adding AVU metadata to that file can trigger a similar policy enforcement point.

2. Addition of a messaging mechanism to pass these events through topics and queues over an asynchronous messaging framework such as RabbitMQ.

3. Development of a standard framework, based on distributed OSGi containers [10] that allows definition of custom indexers that receive these events.

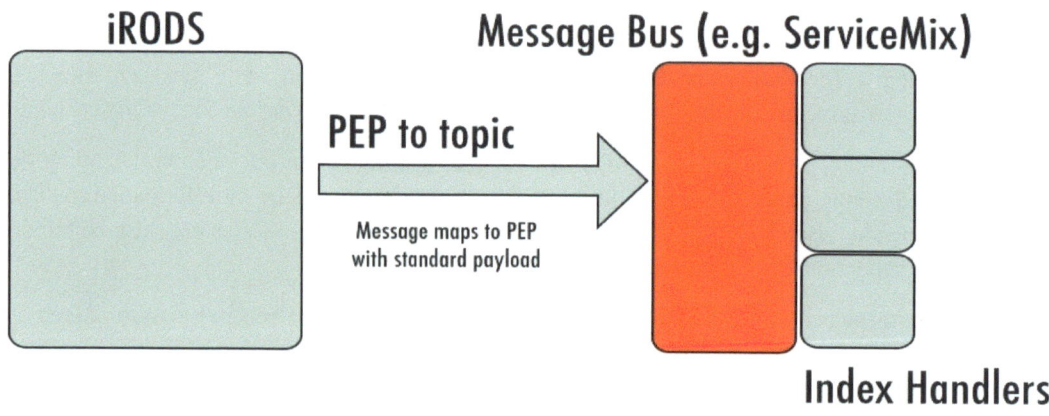

llustration 2: Basic indexer topology

Implementations of indexers can be configured as OSGi components, giving them a distributed, fault-tolerant topology. Indexers can be of two primary types, metadata-only indexers, and file content based indexers. metadata only indexers rely solely on AVUs and related iRODS catalog information. Such indexing events are processed by transforming data from the policy enforcement point into an operation on an external index. In the case of HIVE, adding a SKOS term as a serialized RDF statement about a file or collection causes an event to be sent to a triple-store indexer that can add and update a Jena triple-store with statements about the iRODS collections.

The second type of indexer is a file content indexer. In the case of an indexer that must process all or part of an iRODS file, for subsetting, metadata extraction, inverted index generation, or other purpose, it is necessary to access the actual file contents through some sort of streaming i/o operation. Indexing based on file contents is relatively expensive compared to pure metadata-based indexing, as it requires either streaming data from iRODS, or acting on the data at rest. A second complication is that indexers are independent of each other, and there may be multiple indexers interested in the content of the files. As we discuss computation in a further section, we will be introducing the concept of computational resources, and expand on the basic idea of utilizing specialized iRODS resources that can stage files through copy or replication for indexing by multiple indexers, with additional capability to trim or garbage collect files after indexing is accomplished.

Through either technique, an indexer can plug into event streams originating with iRODS policy enforcement points, and through asynchronous mechanisms that limit pressure on operational grids, provide near real-time indexing synchronization. Using this arrangement, iRODS can focus on policy-managed preservation of data and metadata, serving as a canonical representation, and different parts of the grid may be projected into ephemeral indexes of arbitrary type, to be created, removed, and regenerated as needed.

Virtual Collections

DFC is introducing the concept of 'Virtual Collections" in order to bring indexed metadata stores back into the reach of the data grid. iRODS provides the ability to organize collections through a metadata catalog, giving a global logical namespace that is detached from the underlying physical data stores. The composition of collections, currently, is limited to the arrangement of these collections in the iRODS catalog based on parent/child relationships. Virtual collections are a method to integrate queries of the iRODS catalog or generated external indexes on an equal footing to the primary 'index', which is the iCAT organization of Collections and Data Objects. Virtual collections are defined as any query that can produce a listing equivalent to the "ils -LA" iCommand.

Virtual collections, like metadata templates, are defined through a combination of abstract interfaces and a reference implementation class. There are two primary objects required to implement virtual collections. First is a maintenance service that is responsible for discovering, serializing, and maintaining the definition of a virtual collection. The second is an execution service that can accept parameters and configuration, formulate the appropriate query, and format the results into a listing. The current reference implementation defines virtual queries as JSON structures, decorated with AVU metadata, and stored within iRODS. The current implementation is focused on access via the new DFC web interface, which is in the DFC GitHub repository, and called the "irods-cloud-browser". [11]

Given the ability to form new collections (even across federated zones) based on any arbitrary metadata, is a powerful new capability, and central to the fulfillment of the vision of a collection life-cycle supported by cyber-infrastructure. In summary, iRODS can define structures and validation requirements for metadata stored n the grid. That metadata can be managed and secured for the long term via the application of preservation policies. This metadata can be projected into an arbitrary arrangement of indexes, either based on metadata in the catalog, or based on the contents of data objects, processed through any available algorithm. These indexes, once created, can then be used to discover and arrange new collections, across federations, and orthogonal to any folder arrangement within iRODS.

Concept Summary

The DFC approach to discovery and collection formulation is mindful of the requirements of cyber-infrastructure, and enables the fulfillment of several aspects of Moore's life-cycle approach to research collections. The capabilities being pioneered in DFC are being developed so as to easily flow back into the iRODS community. Using the policy-managed approach of iRODS, DFC is developing metadata curation facilities, enhanced indexing facilities, and the ability to integrate external indexes back into grid operations.

Tools for Computation

IRODS has traditionally focused on policy management of scientific data. [12] In the original context, this approach identified the concept of procedures, which are encapsulated in micro-services. These procedures are then activated using policies that are implemented at policy enforcement points (PEPs). The goal of these policies is to mediate security, auditing, preservation, metadata extraction, and other data management concerns. Under DFC, the concept of procedures is expanding beyond the original micro-services construct, and taking on a more general notion of computation acting on data. Under DFC, a capability called 'Workflow Structured Objects' (WSO) has been demonstrated. These WSOs encapsulate scientific workflows, including input parameters, workflow processes, and outputs, such that execution of a WSO given a set of input parameters preserves those parameters and inputs, documenting a process, and enabling reproducible research. [13] WSOs demonstrate a broad set of new approaches that have been developed within DFC, and among DFC collaborators similarly utilizing iRODS. That is, the confluence of computation and data management under the policy-managed approach of iRODS.

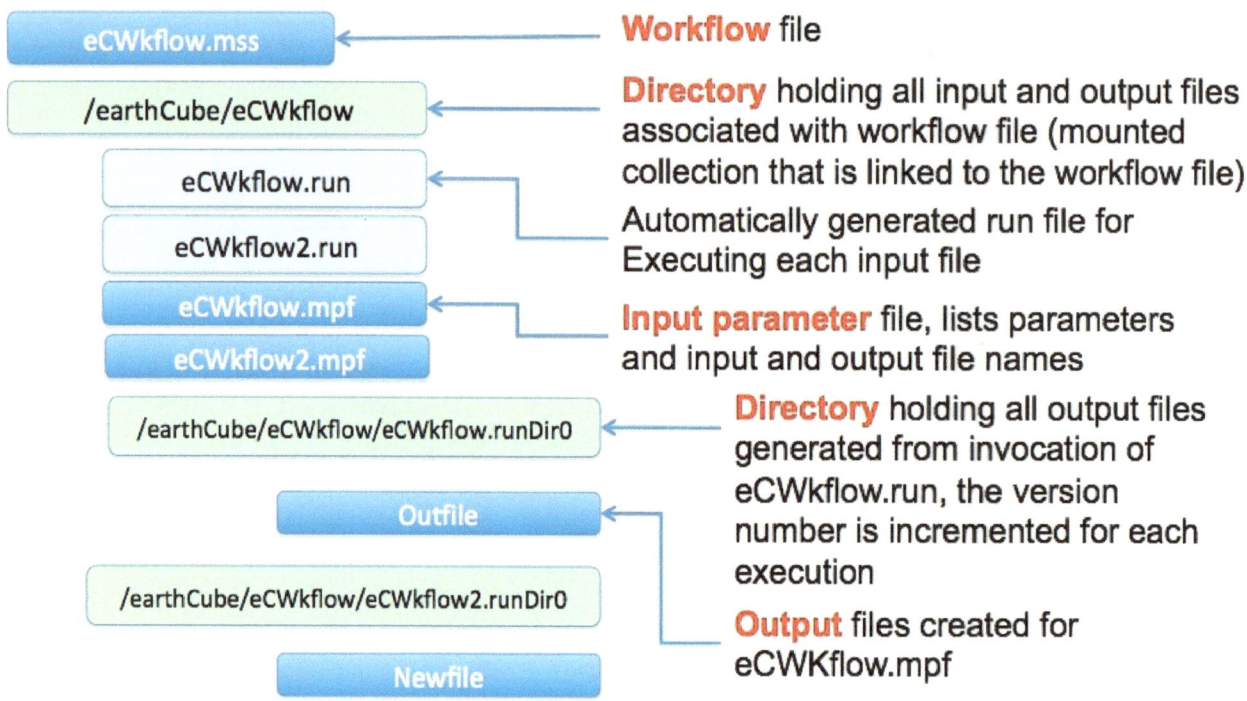

Illustration 1: Workflow Strucured Objects (Rajaseka 2014)

Computation merged with data management under the umbrella of policy-management is important in several respects.

1. As a base capability in Gannon's cyber-infrastructure concept, the ability to manage computation in the same context as the underlying data is an affordance to researchers. Data can be processed as easily as it can be manipulated and copied for data management purposes.

2. Links between computation and data provide a basis for provenance tracking. As data is processed, the derivative products may be registered back into the grid environment, along with metadata describing the workflows, parameters, and processing details. By providing computational affordances, the DFC cyber-infrastructure receives important audit trail and provenance information in return.

3. The computational processes that have been used to create new data products are metadata, describing the significant properties of a data object, and helping to ensure that the data object remains useful to a designated community over the long term. Thus, the merging of computation and data management contributes to the long term preservation goals, and ensures the viability of reference collections and published data sets at the mature end of the research data life-cycle.

Rajaseka points out an emerging picture of computation as it pertains to DFC and iRODS. He describes a paradigm shift in several respects. [13]

1. A shift from compute-intensive processing to data-intensive processing. The data and data handling become a primary focus of computational processes.

2. A shift from "large actions on small data" to "small actions on large data", meaning that the costs of data movement is beginning to outstrip the costs of the computational processes themselves.

3. From "move data to the processing site" to "move process to the data site". Discovery of appropriate locations to do computation on data, as well as facilities to host and schedule such computation become pressing issues.

4. "Function chaining" to "service chaining". Computation becomes a modular process, were linked capabilities may be used to accomplish a computational tasks.

5. "Model-based science" to "data-based science". Data mining and computation on large sets of observations is becoming an important practice in data-driven science.

WSOs are but one example of DFC cyber-infrastructure innovation on the iRODS platform. Other groups that participate in DFC have reached similar conclusions and have begun developing similar approaches, notably the iPlant Collaborative, who have made great advances in marrying computation with data management over the iRODS platform. [14] The discovery environment is a workbench for researchers that manages data access, provides an abstraction of applications for data analysis, and a facility for managing the data products and notifications from these computational tasks. The iPlant foundation APIs are REST accessible services that can serve as a template for formalization and further abstraction within the DFC environment, a task that is currently in progress.

1. iPlant provides an abstraction for computation as an application, with metadata describing its purpose and parameters, such that users can discover and interact with these applications.

2. Gateways have been created to various workflow and computation environments, including Condor, Docker containers, and grid resources such as Exceed.

3. A facility to monitor execution and receive notifications of progress and completion of computational processes has been created.

There are several subsystems and general features here that are being incorporated into the DFC architecture, through implementation of the Discovery Environment software on the DFC grid, and further enhancement of iRODS capabilities. For example, the DFC design contemplates extending the 'application' abstraction demonstrated in iPlant to include WSOs and iRODS user submitted rules. This provides a basis for interfaces for users that allow each grid to extend its capability by providing the appropriate procedures and computational resources.

An especially interesting development in computing is the emergence of lightweight containers such as Docker. [15] Docker provides a means to define containers consisting of a base operating system and applications, as configuration, and to rapidly distribute and launch containers for computation. Under discussion in DFC is the concept of a computational resource, which is a specialized resource that could manage Docker based applications, collocated with an iRODS storage resource This arrangement would allow many "small actions on large data" operations to take place on a platform that also contains the target data resources. Such containers could be tied into the 'application' constructs described above, and illustrated by the current Discovery Environment services.

It is important to note that the distinction between computation and policy management of data is actually quite artificial. Services that have been traditionally considered data-management, such as metadata extraction or file format translation, can be mediated through the policy management framework of iRODS can actually take place using the same application constructs that are employed in the Discovery Environment. This is especially the case if we consider the indexing framework. Earlier, we described a use case for indexers of file contents, such as text indexing, file format recognition, and data subsetting for metadata extraction. These use cases are, in essence, computation on data, potentially generating new data products. In fact, the previously mentioned metadata templates contemplate the ability to declaratively describe automatic metadata extraction as an 'application', and this application could be accomplished in the described framework.

These use cases employ a common set of services that define new iRODS components and capabilities driven by DFC development.

The policy-managed approach of iRODS, mediated through policy enforcement points, and through the policies and procedures invoked by the iRODS rule engine, has proven to be a successful data management paradigm. It is worthwhile, in the context of DFC development, to extend these concepts to the management of computation as it intersects with research data. It is true that the iRODS based cyber-infrastructure cannot cover every single use case, but it is arguable that there are distinct classes of cases where grid users need to discover and apply computation to data on the grid, to be notified of the status of this computation, to be able to manage the products of this computation. As cyber-infrastructure, DFC needs to automatically preserve metadata about provenance, history, and meaning of derived data products. Existing implementations at DFC partners such as iPlant demonstrate the viability of these approaches at scale. Computational facilities may be common across preservation and data management use cases as well as data driven research use cases. IRODS consortium efforts to standardize plug-ability at many levels, including at the level of the rule engine, to develop compose-able resource hierarchies, and to provide grid introspection and discovery services, all lead towards an emerging set of capabilities where computation and data management meet.

[4] CONCLUSION

The development of DFC architecture is resulting in many novel enhancements to iRODS in support of key requirements of cyber-infrastructure. In particular, innovations in the area of metadata management and discovery, and in the application of computation to grid-managed data constitute significant new capabilities for iRODS. These discovery and computational features are necessary in order for DFC to achieve its goal of providing support the entire life-cycle of research data.

[5] ACKNOWLEDGMENTS

The development of the iRODS data grid and the research results in this paper were funded by the NSF OCI-1032732 grant, "SDCI Data Improvement: Improvement and Sustainability of iRODS Data Grid Software for Multi-Disciplinary Community Driven Application," (2010-2013), and the NSF Cooperative Agreement OCI-094084, "DataNet Federation Consortium", (2011-2013).

[6] REFERENCES

[1] "DataNet Federation Consortium Proposal" [Online]. Available: `http://datafed.org/dev/wp-content/uploads/2012/04/DFCproposal.pdf`. [Accessed: 30-May-2015].

[2] "iRODS Overview" [Online]. Available: http://irods.org/wp-content/uploads/2012/04/iRODS-Overview-November-2014.pdf. [Accessed: 30-May-2015].

[3] "Virtualization of the data life cycle". R. Moore [Online]. Available: *https://conferences.tdl.org/tcdl/index.php/TCDL/2010/paper/view/97*. [Accessed: 30-May-2015].

[4] D. Gannon, B. Plale, M. Christie, Y. Huang, S. Jensen, N. Liu, S. Marru, S. L. Pallickara, S. Perera, S. Shirasuna, and others, "Building Grid portals for e-Science: A service oriented architecture," *High Perform. Comput. Grids Action*, 2007.

[5] "imeta - IRODS," 29-May-2015. [Online]. Available: https://wiki.irods.org/index.php/imeta. [Accessed: 29-May-2015].

[6] "DICE-UNC/jargon-extensions-if · GitHub," 29-May-2015. [Online]. Available: https://github.com/DICE-UNC/jargon-extensions-if. [Accessed: 29-May-2015].

[7] "DICE-UNC/jargon-extensions · GitHub," 29-May-2015. [Online]. Available: https://github.com/DICE-UNC/jargon-extensions. [Accessed: 29-May-2015].

[8] J, Greenberg et al, "HIVEing Across the U.S. DataNets" [Online]. Available: https://rdmi.uchicago.edu/sites/rdmi.uchicago.edu/files/uploads/Greenberg,%20J,%20et%20al_HIVEing%20Across%20the%20U.S.%20DataNets_0.pdf. [Accessed: 29-May-2015].

[9] "Rules - IRODS," 29-May-2015. [Online]. Available: https://wiki.irods.org/index.php/Rules. [Accessed: 29-May-2015].

[10] "OSGi Alliance | Main / OSGi Alliance." [Online]. Available: http://www.osgi.org/Main/HomePage. [Accessed: 30-May-2015].

[11] "DICE-UNC/irods-cloud-browser · GitHub." [Online]. Available: https://github.com/DICE-UNC/irods-cloud-browser. [Accessed: 30-May-2015].

[12] R. Moore, A. Rajasekar, and M. Wan, "Policy-Based Data Management," in *IEEE Policy 2010 workshop, George Mason University*, 2010.

[13] A. Rajasekar, "iRODS Workflows and Beyond" [Online]. Available: http://irods.org/wp-content/uploads/2014/06/Workflows-iRUGM-2014.pdf. [Accessed: 30-May-2015].

[14] "The iPlant Collaborative: Cyberinfrastructure for Plant Biology." [Online]. Available: http://www.ncbi.nlm.nih.gov/pmc/articles/PMC3355756/. [Accessed: 31-May-2015].

[15] "Docker - Build, Ship, and Run Any App, Anywhere." [Online]. Available: https://www.docker.com/. [Accessed: 31-May-2015].

A Flexible File Sharing Mechanism for iRODS

Alva Couch
Tufts University
Medford, MA
couch@cs.tufts.edu

David Tarboton
Utah State University
Logan, UT
dtarb@usu.edu

Ray Idaszak
Rennaisance Computing
Institute (RENCI)
Chapel Hill, NC
rayi@renci.org

Jeff Horsburgh
Utah State University
Logan, UT
jeff.horsburgh@usu.edu

Hong Yi
Rennaisance Computing
Institute (RENCI)
Chapel Hill, NC
hongyi@renci.org

Michael Stealey
Rennaisance Computing
Institute (RENCI)
Chapel Hill, NC
stealey@renci.org

ABSTRACT

The traditional iRODS mechanisms for file sharing, including user groups, often require some form of iRODS administrative privilege. In the HydroShare project for enabling hydrology research, we perceived a need for more flexible file sharing, including unprivileged creation and management of user groups according to policies quite distinct from the Linux/Unix policies that initially motivated iRODS protections. This is enabled by a policy database in PostgreSQL and management API written in Python that are deployed in parallel to iCAT. Innovations in iRODS 4.1 allow access control based upon this PostgreSQL database rather than the default iCAT server, by interposing access control code before the access event using iRODS Policy Enforcement Points. The result is an access control mechanism that closely matches scientific needs for file sharing, and brings "dropbox-like" file sharing semantics to the network filesystem level.

Keywords

File sharing, authorization, access control, policy enforcement points

INTRODUCTION

The access control mechanisms implemented by iCAT for iRODS [1, 2, 3] – while suitable for a variety of types of file sharing – fell short of our requirements for file sharing in the data-centered social networking site "HydroShare" [4, 5, 6] (http://www.hydroshare.org). Based upon the Hydrologic Information System (HIS) [7] of the Consortium of Universities for the Advancement of Hydrologic Science, Inc. (CUAHSI), HydroShare brings social networking to data centered science, by enabling "object-centered" discussion in data-enabled science [8]. HydroShare enables posting of a large variety of data types as directories of files in the BAGIT [9] format, and stores metadata for each object in the object, using the Open Archives Initiative metadata guidelines [10] tailored to the needs of water sciences. The goal of HydroShare is to enable sharing of hydrologic data and models with the same ease with which one stores photos or videos on social networking sites, with social mechanisms including commenting, rating and sharing of data objects.

iRODS does not currently provide a sufficient protection model for objects in HydroShare, because of our desire to enable social features rather than just file access. Among other issues, iRODS normally requires administrative access in order to:

1. Change the owner of a file.

2. Create and manage a user group.

iRODS UGM 2015 June 10-11, 2015, Chapel Hill, NC

We desired a more flexible concept of object ownership, sharing, and mutability in which a user without administrative access can:

1. Reassign ownership and all privileges over an object to another user.

2. Create, manage, and destroy a user group and its privileges.

The result was the "IrodsShare" project, a sub-project of the "HydroShare" project, centered specifically upon creating access control mechanisms more suitable for the collaborative scientific data sharing desired in HydroShare than the default mechanisms in iRODS. We expect that this sharing model developed for HydroShare will be of broader interest for scientific data sharing beyond hydrology and water science.

HydroShare

The "HydroShare" project [4, 5, 6] aims to create social mechanisms for data sharing between researchers in the water sciences, that enhance the value of the data to other researchers. A HydroShare "resource" is a bag [9] of data that is accessible as one unit, and can contain many subfiles of potentially differing formats. It is best to think of a HydroShare "resource" as a directory of files rather than a single file, though single file resources are easily represented in bags. This use of the word "resource" is completely detached from the iRODS meaning of the word "resource"; HydroShare "resources" are directories adhering to a strict format and requirements for contents and metadata.

HydroShare is written in Python/Django and utilizes iRODS as a data management back end. Currently, HydroShare functions according to a Django access control model and allows or denies access via the HydroShare website. It is desirable to create a coherent access control mechanism that is homogeneous between the privileges granted at the Django website and all forms of access to HydroShare resources via iRODS, including REST, iCommands, etc. "IrodsShare" provides this desired mechanism.

Policies for file access

Our policies for file control are somewhat different than those in typical iRODS or Linux, and aligned around social and scientific needs and priorities rather than filesystem traditions. All objects in HydroShare are "HydroShare resources": bags of data containing perhaps multiple constituent files. The file protection model is very simple:

1. A user can possess "View", "Change", or "Owner" privileges over a specific resource.

2. "Owner"s can change any aspect of a resource, including its title, ownership, and access to others, and can delete the resource.

3. "Change" users are limited to changing the resource contents and/or metadata.

4. "View" users are limited to viewing the contents of the resource file.

5. Resources can have multiple "Owner"s.

6. The last "Owner" of a file cannot revoke personal ownership; ownership must be assigned to another user first.

So far, these privileges should be familiar to most users. However, the system diverges from standard file access due to several extra boolean resource attributes, including:

1. "Public" – the content of this resource is made available to all users.

2. "Discoverable" – the existence and metadata for this resource file is available to all users, but not necessarily its content.

3. "Shareable" – this resource can be shared with other users and groups by a non-owner.

These settings are used to manage whether "IrodsShare"s flexible sharing mechanisms are enabled for a specific resource. A maximally protected resource in HydroShare would have "Shareable", "Discoverable", and "Public" set to FALSE. This is a resource that an owner can share with selected users, but that non-owners cannot in turn share with others. This is the typical setting for resources that contain pre-publication data.

1. Setting "Shareable" to TRUE allows people and groups with whom the resource has been shared to re-share it with others. This – in essence – enables limited distribution of the resource via transitive sharing. For example, a researcher can share the resource with that researcher's research assistants, and those assistants – in turn – can share it with anyone they wish.

2. Setting "Discoverable" to TRUE allows researchers not associated with the owners to learn of the existence of the resource and request access from the owners. This enables the owners of the resource to reach a common understanding of the limits of use with the requester before sharing the resource. The use of this flag is to enable pre-publication access to data on a limited basis by trusted external researchers.

3. Setting "Public" to TRUE enables general access to the resource, by any user, and is normally enabled after results are published. When this flag is TRUE, any user anywhere can discover and download the data.

The default in HydroShare is to set "Shareable" to TRUE and the others to FALSE.

Sharing and groups

Perhaps the most unique aspect of IrodsShare – from the perspective of resourcesystem-like protections – is the concept of resource "Sharing". Any user with a privilege over a "Shareable" resource can grant that privilege to others. As well, privileges previously granted by a specific user can be removed by that user. One can share a resource with either another user or a group of users.

In like manner, user groups can be created, managed, and destroyed at will by any user. A group of users is managed very much like a resource in IrodsShare:

1. Any user can create and thus own a group.

2. Users can have "View", "Change", or "Owner" privilege over a group.

3. "Owner"s can invite and remove group members (with any privilege) or destroy the group.

4. "Change" users can invite new members to a "Shareable" group, and can assign those members "Change" or "View" privileges over the group.

5. "View" users cannot invite new members, even to a "Shareable" group.

As with resources, a group can be "Shareable" or not.

USE CASES

This model was conceived based on many conversations and experience working with water scientists interacting with the CUAHSI Hydrologic Information System implemented at the CUAHSI Water Data Center. It strives to balance the needs and concerns surrounding sharing and enabling access to scientific data while protecting the rights and desires of the researchers involved. These users are a set of water scientists with broad interests, from basic hydrology to water quality and various forms of modeling. We discussed what scientists really need, instead of limiting discussion to resource sharing mechanisms that currently exist.

The resulting design was documented in a working policy document [11] and motivated by several scientific use cases discussed below. As well, part of the design attempts to provide the basic infrastructure to become a DataOne node [12] and to serve the sharing needs of water science groups such as the Critical Zone Observatories, as documented in [13].

In the following, it is important to remember that HydroShare "resources" are actually – logically – directories of related files. Thus access to a resource means access to that directory and all files contained therein.

There are many non-conflicting interpretations of a HydroShare resource, including:

1. Files for one trial of a single "experiment".

2. Input, output, and software for a specific model run.

3. A set of related files that represents the published data output of a project.

and many other potential contents.

Like a UNIX group, the concept of an IrodsShare group has many non-conflicting interpretations, including:

1. Users associated with a specific project.

2. Users associated with a specific organization or laboratory.

3. Co-authors of a specific academic paper.

4. A loosely managed group of users sharing the same interests.

While the first and second interpretations apply to a regular user group in Linux or iRODS, the third and fourth interpretations are only really practical in IrodsShare, because of the administrative overhead involved in finding an iRODS administrator to create what is really an "ad hoc" notion of group.

The word "group" does not serve particularly well to describe the IrodsShare group concept because of its common association with the narrower definition used in iRODS and linux file systems. However, no better descriptive word has emerged.

Researcher plus graduate students and collaborators

To start a project, the researcher creates a group and assigns graduate students to the group with appropriate privilege (typically "View") (left side of Figure 1). As students enter and leave the project, their group access is managed by the researcher. A senior graduate student can be delegated the task of managing group membership by assigning group "Change" privilege to that student.

Next, the researcher shares resources at different privileges with group members. In this example (right side of Figure 1), the whole group has "View" privilege while "assistant 1" has "Change" privilege. The access control system automatically defaults to the highest level of privilege granted through any mechanism.

Thus, the researcher retains control and oversight over all project resources and can delegate responsibility to students very easily. Most notably, students with change privilege do not automatically assume ownership of a resource after they make changes. Each resource remains owned by the original creator – the researcher. This eliminates the very common plight that graduate students acquire too much privilege, and then graduate.

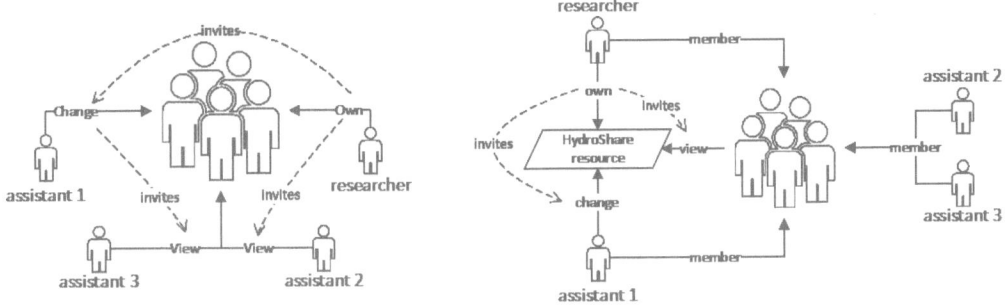

Figure 1. User groups and resource sharing in IrodsShare.

Pre-publication and post-publication data protection

The "Shareable" flag on each resource arose from the necessity of protecting resources from dissemination before results are published. The researcher can set this flag to FALSE so that no non-owner can share an accessible resource with another user or group.

For example, this is desirable if details of the original study that collected the data in question have not been published. Then the owner – e.g., an academic researcher – can delegate "Change" privilege to students without permitting students to disseminate the data to others. Likewise, the "Discoverable" flag arose from the desire to enable and encourage polite and acknowledged reuse of pre-publication data by permission of the owner. Setting this flag to TRUE allows all users to discover that the resource exists, but users must then request access from an owner, typically via email in this implementation. This allows the owner to set limits on resource use.

Finally, the "Public" flag arose from the desire to be able to publish results for general consumption after the results of the data have been published. Setting this flag to TRUE gives general access to the data to all users.

Miscellaneous use cases

A few other features enable other scientific needs, including:

1. "Published" – a resource flag that indicates that the resource has been assigned a Digital Object Identifier (DOI).

2. "Immutable" – a resource flag that indicates that the owner has frozen the contents of a resource, overriding "Change" privileges for all current users holding that privilege.

POTENTIAL DESIGN APPROACHES

To accomplish these requirements, we discussed and tried several design approaches before settling on the final approach. Ironically, while the project was to add non-admin functionality to iRODS, several approaches were discarded because the unprivileged user is *too privileged* and, at the present time, there is no formal access control on certain aspects of iCAT, including Attribute-Value Units (AVUs).

For example, we considered adding access control to iCAT metadata Attribute-Value Units (AVUs). Since there is no access control on AVUs, however, we would have had to create that access control. We considered writing several Policy Enforcement Point services to limit the changes one can make to access control AVUs. We abandoned this approach due to four main reasons:

1. The code for controlling AVU content was discovered to be very complex.

2. The code for utilizing AVU content for access control was predicted to be unacceptably slow.

3. Other projects' use of AVUs could potentially collide with our uses.

4. iRODS Policy Enforcement Points had not been used previously for this function and it was not clear at the time whether this would work properly.

Thus, after some prototyping and proof of concept, we concluded that this approach was more complex than warranted.

Other considered approaches included modifying the iCAT itself to include a different kind of metadata. This was abandoned because it was simpler for us to maintain a separate database just for access control than to branch the iCAT code.

THE CHOSEN DESIGN

After consultation with the iRODS team at RENCI, we decided to base our access control system upon a new feature of iRODS dynamic Policy Enforcement Points (PEPs) that was implemented for us and will be supported in the iRODS 4.1 release. If a function defined for a pre-execution PEP returns an error, then the action being protected by the PEP is canceled. Thus, we define PEP code for pre-execution before resource creation, update, and read requests, and this suitably enforces our access control by returning "Access Denied" errors as appropriate. This code reads an external PostgreSQL database to determine user privileges (Figure 2). In this figure, arrows represent directions of control during data access and policy enforcement. The API controls access policy, which in turn controls iRODS resource access.

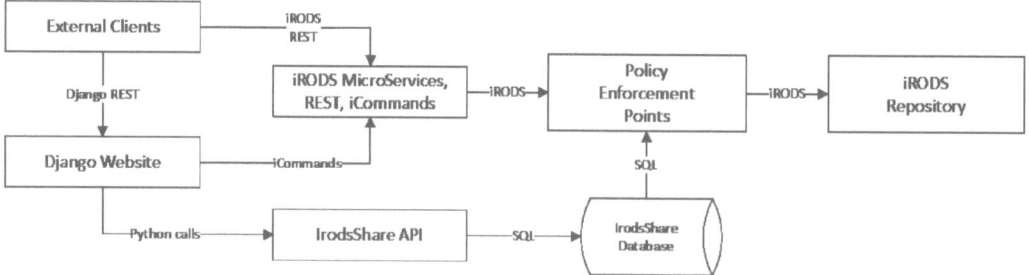

Figure 2. The access control mechanism interposes policy enforcement points between iRODS resources and a user website, thus providing consistent access regardless of the access mechanism in use.

In order for this to work, all resources being protected in this fashion must be world-writeable according to each user's iRODS protections. We accomplish this by making all resources writeable to a designated "HydroShare" iRODS group, in which all users are members. Thus, the access control scheme is subtractive; it denies access to resources to which the iRODS user would otherwise have access according to normal iRODS protections. Because of this implementation, it is unimportant whether the actual resources are owned by specific iRODS users or not; ownership in iRODS is orthogonal to ownership in IrodsShare, and we are currently using a proxy user as the iRODS owner.

The IrodsShare API

The contents of the access control database are managed by a Python API (to interact with the Django/Python website, and to suit the maintenance requirements of the HydroShare team). The API requires authentication of a HydroShare user and acts on behalf of that user. All policies regarding access control are implemented via the API, which modifies the PostgreSQL access control database directly. In turn, this database is shared with iRODS microservices to control the PEPs as discussed above.

While – ideally – we would like this to be accessible via iRODS REST microservices, that feature has not yet been implemented.

Sharing Policies

In any social sharing situation, there is a balance between enabling sharing and minimizing "SPAM" or other unwanted information. This concern complicates operating policies somewhat, because sharing can in principle be a source of SPAM. For example,

1. A user must give permission to be made an owner of a resource. This prevents users being inadvertently made responsible for resources without their knowledge.

2. A user must give permission to join a group, after being invited by an authorized member. This prevents users from placing everyone in a group, for the purpose of broadcasting information to everyone.

Sharing with individual users is otherwise unmoderated, and anything shared with a group is immediately distributed to its members.

PROJECT STATUS

At present, only the database and Python policy engine are complete and tested, with documentation in the Sphinx documentation system. Deployment to HydroShare has been deferred due to the need to deal with issues in Django access control first. PEP enforcement has been designed but not implemented.

CONCLUSIONS

We presented a novel model of access control for scientific data that mimics file sharing in social networks, at the filesystem level. This mechanism strikes a balance between capability at the filesystem level and ease of implementation, by strategic use of iRODS dynamic Policy Access Points. By setting iRODS to "allow anything" and then constructing selective denials, we change as little of iRODS as possible to accomplish this.

This is just a first step toward moving social networking to the filesystem level. While critics of IrodsShare believe that it creates a two-phase commit problem with Django, this criticism is only valid as long as Django and iRODS are competing to be ground truth for access control.

As for the actual usability and utility of the project, only time and experience will tell. We believe we have struck a balance between utility, ease of understanding, and limitation of abuse potential.

Availability

The project is freely available from http://github.com/hydroshare/IrodsShare.

ACKNOWLEDGMENTS

This work was supported by the National Science Foundation under collaborative grants ACI-1148453 and ACI-1148090 for the development of HydroShare (http://www.hydroshare.org). Any opinions, findings and conclusions or recommendations expressed in this material are those of the authors and do not necessarily reflect the views of the National Science Foundation.

We are also particularly indebted to the scientists of the HydroShare project, who vetted our access control design carefully at repeated meetings over more than two years. The iRODS team at RENCI – and particularly Jason Coposky and Antoine de Torcy – provided crucial help in evaluating options for implementing this software, and implemented features of the Python library for iRODS specific to this project.

REFERENCES

[1] Reagan Moore, "Toward a Theory of Digital Preservation", *International Journal of Digital Curation* 3(1), June 2008.

[2] Terrell Russell, Jason Coposky, and Michael Stealey "Hardening iRODS for an Initial Enterprise Release (E-iRODS)" *Proc. 2012 iRODS User Group Meeting,* Renaissance Computing Institute (RENCI), Chapel Hill, NC.

[3] Terrell Russell, Jason Coposky, Harry Johnson, Ray Idaszak, and Charles Schmitt, "E-iRODS Composable Resources", *Proc. 2013 iRODS User Group Meeting,* Renaissance Computing Institute (RENCI), Chapel Hill, NC.

[4] Tarboton, D. G., R. Idaszak, J. S. Horsburgh, J. Heard, D. Ames, J. L. Goodall, L. E. Band, V. Merwade, A. Couch, J. Arrigo, R. Hooper, D. Valentine and D. Maidment, "A Resource Centric Approach for Advancing Collaboration Through Hydrologic Data and Model Sharing," *Proceedings of the 11th International Conference on Hydroinformatics (HIC 2014)*, New York City, USA, http://www.hic2014.org/proceedings/handle/123456789/1539.

[5] Tarboton, D. G., R. Idaszak, J. S. Horsburgh, J. Heard, D. Ames, J. L. Goodall, L. Band, V. Merwade, A. Couch, J. Arrigo, R. Hooper, D. Valentine and D. Maidment, (2014), "HydroShare: Advancing Collaboration through Hydrologic Data and Model Sharing," in D. P. Ames, N. W. T. Quinn and A. E. Rizzoli (eds), *Proceedings of the 7th International Congress on Environmental Modelling and Software,* San Diego, California, USA, International Environmental Modelling and Software Society (iEMSs), ISBN: 978-88-9035-744-2, http://www.iemss.org/sites/iemss2014/papers/iemss2014_submission_243.pdf.

[6] Heard, J., D. Tarboon, R. Idaszak, J. Horsburgh, D. Ames, A. Bedig, A. M. Castronova, A. Couch, P. Dash, C. Frisby, T. Gan, J. Goodall, S. Jackson, S. Livingston, D. Maidment, N. Martin, B. Miles, S. Mills, J. Sadler, D. Valentine and L. Zhao, (2014), "An Architectural Overview of Hydroshare, A Next-Generation Hydrologic Information System," *Proc. 11th International Conference on Hydroinformatics (HIC 2014)*, New York City, USA, http://www.hic2014.org/proceedings/handle/123456789/1536.

[7] Tarboton, D.G., Horsburgh, J.S., Maidment, D.R., Whiteaker, T., Zaslavsky, I., Piasecki, M., Goodall, J., Valentine, D., Whitenack, T. (2009). "Development of a community Hydrologic Information System." In: Anderssen, R. S., R. D. Braddock, and L.T.H. Newham (eds.) *18th World IMACS Congress and MODSIM09 International Congress on Modelling and Simulation,* Modelling and Simulation Society of Australia and New Zealand and International Association for Mathematics and Computers in Simulation, July 2009, pp. 988-994, ISBN: 978-0-9758400-7-8.

[8] Engestrm, J. (2005). *Why some social network services work and others dont Or: the case for object-centered sociality,* http://www.zengestrom.com/blog/2005/04/why-some-social-network-services-work-and-others-dont-or-the-case-for-object-centered-sociality.html, last accessed 5/9/2015.

[9] Boyko, A., J. Kunze, J. Littman, L. Madden, B. Vargas (2012). *The BagIt File Packaging Format (v0.97),* (Network Working Group Internet Draft), available at http://tools.ietf.org/html/draft-kunze-bagit-10, last accessed 2/20/2015.

[10] Lagoze, C., Van de Sompel, H., Johnston, P., Nelson, M., Sanderson, R., Warner, S. (2008a). *Open Archives Initiative Object Reuse and Exchange: ORE Specification Abstract Data Model,* available at http://www.openarchives.org/ore/1.0/datamodel.html, last accessed 3/3/2015.

[11] Alva Couch and David Tarboton, *Access Control for HydroShare,* internal project document, verson as of May 1, 2015.

[12] DataONE (2015). *DataONE Architecture, Version 1.2,* available at http://releases.dataone.org/online/api-documentation-v1.2.0/, last accessed 3/7/2015.

[13] Zaslavsky, I., Whitenack, T., Williams, M., Tarboton, D.G., Schreuders, K., Aufdenkampe, A. (2011). "The initial design of data sharing infrastructure for the Critical Zone Observatory," *Proceedings of the Environmental Information Management Conference,* Santa Barbara, CA, 28-29 September, EIM 2011, http://dx.doi.org/10.5060/D2NC5Z4X.

Resource Driver for EMC Isilon: implementation details and performance evaluation

Andrey Nevolin
EMC
EMC Russia&CIS, Begovaya st., 3/1, 30 floor, Moscow, Russia, 125284
andnev@mail.ru

Andrey Pakhomov
EMC
EMC Russia&CIS, Begovaya st., 3/1, 30 floor, Moscow, Russia, 125284
Andrey.Pakhomov@emc.com

Stephen Worth
EMC
62 T.W. Alexander Dr.
Durham, NC 27709
Stephen.Worth@emc.com

ABSTRACT

We developed a resource driver that provides iRODS with the full potential of an EMC Isilon scale-out storage system. Previously Isilon connected to iRODS as a legacy NFS device. Our dedicated Isilon resource driver offers Kerberos authentication, better read performance, and load distribution among nodes of a clustered storage system. Also 'intelligent' storage system features may be potentially accessed from a dedicated resource driver. All of the listed advantages cannot be achieved in the context of an NFS usage model (except Kerberos, which is not common for NFS environments). In this paper we describe the motivation behind the dedicated Isilon driver, implementation details, and results of initial performance evaluations of the driver. Initial performance data shows that the driver provides 30%-better read performance and almost perfect load balancing between the nodes of an EMC Isilon storage system (a clustered storage system).

Keywords

iRODS, EMC, Isilon, HDFS, NFS, Resource Driver.

INTRODUCTION

EMC Isilon scale-out storage system is popular for storing large amounts of unstructured data. This is especially true for the Life Sciences domain [1]. iRODS[1] (The Rule Oriented Data System) in its turn is a popular solution for managing unstructured data [1]. Both Isilon and iRODS have been used together for several years in different organizations for storing and taking control of multiple terabytes and even petabytes of data [1].

Till now Isilon has been represented to iRODS as legacy NFS storage leveraging the '**unixfilesystem**' iRODS resource driver type. While this approach suits well for many purposes, it doesn't align well with modern data management practices. Below is a list of reasons demonstrating why NFS[2] is not a perfect option for accessing Isilon:

1) **Mounting NFS devices requires super-user privileges.** This complicates the management of a data grid system like iRODS. Super-user intervention is required for trouble-shouting, initial configuration, and re-

---

[1] All information in this article is relevant with respect to iRODS 4.0.3 version, which is the latest version at the time of article creation

[2] Here and everywhere in the document 'standard' NFS client implementation – that comes with UNIX distributions – is implied

iRODS UGM 2015, June 10-11, 2015, Chapel Hill, NC.

configuration. That can be a bigger problem in cases where the roles of machine administrator and Data Grid administrator belong to different people

2) **Load balancing between different storage system's nodes is problematic.** Typically each iRODS storage resource is associated with a single iRODS Resource server. With respect to NFS that means that all I/O (input/output) workload associated with a storage resource will be processed through a single mount. In its turn, each mount can be served by a single Isilon node only. Isilon is capable of moving mount points from one node to another for load-balancing purposes. But this balancing occurs at the level of network connections, which may be insufficient in cases with very 'busy' connections[3]

3) **Object-style access is not supported by NFS.** Work in object style is becoming popular [4] today. Working with object stores is quite different from conventional file system access because of additional limitations on concurrent access, object modifications, and more.

4) **NFS client is implemented at OS kernel level and is intended for general-purpose usage.** User-space clients may provide better performance, better control of data transfer and may be optimized for particular workloads and environments

5) **NFS doesn't provide access to 'intelligent' features of modern storage systems.**

In our initial design of Isilon resource driver we addressed the first four of the above concerns. Advanced Isilon features may be supported in subsequent releases.

During this work on the driver we identified a number of iRODS issues (24) related to data transfer which were reported. The authors wish to thank the iRODS Community for addressing many of them. They may be found under nickname 'AndreyNevolin'[4] on the iRODS GitHub [5].

IMPLEMENTATION DETAILS

HDFS access to Isilon is the core of our plugin design. HDFS protocol is supported by Isilon since version 6.5 of Isilon OneFS file system [6].

We use low-level HDFS client library Hadoofus [7] to access the HDFS server-side functions. Hadoofus is a very simple library. It implements only client-side counterparts of the server-side API. Relying on the low-level client library allows us to move all data-transfer optimizations to the level of the iRODS resource driver. The driver in its turn is 'iRODS-aware', which allows us to implement iRODS-specific optimizations (and also provide iRODS-specific data-transfer controls available to a user or to the iRODS framework itself). Currently we implemented only simple data pre-fetch and write bufferization. Other optimizations may appear in future. The non-constraining design will allow us to introduce them easily in the future.

Hadoofus is a native C library. It does not rely on the Java runtime environment (in contrast to most HDFS clients). That alleviates concerns related to JRE performance.

Since both Hadoofus and our plugin work in user-space, super-user privileges are not required for creating specialized Isilon resources in iRODS.

HDFS by its nature is an object-based protocol. Our resource driver naturally inherits this property.

---

[3] Consider the following hypothetical example. Imagine a storage system cluster of two nodes. Let's assume next that two NFS connections utilize these nodes: each connection resides on a separate node and each connection consumes 70% of single-node throughput. If one needs to occupy the cluster with a third connection that requires 50% of node throughput for best performance, there is no way to do that without sacrificing performance of two workloads. Even though the cluster still has unallocated throughput of 60% of a single node (30% on each node), this bandwidth cannot be allocated to a single connection

[4] The following link may be used to find all iRODS issues reported by us: https://github.com/irods/irods/issues/created_by/AndreyNevolin

As was stated above, two basic optimizations are currently implemented in the driver to support effective data transfers. The First optimization is data pre-fetch. Data is pre-fetched in blocks of fixed size. When iRODS requests data that is not in a pre-fetch buffer, the resource driver first fills the buffer completely starting from the byte that corresponds to the requested offset. Only after that will the requested data be returned to iRODS. An independent read buffer is associated with each iRODS object opened through the specialized Isilon resource driver[5].

Our second optimization is write bufferization, which is very simple. Data is committed to Isilon only when the client-side intermediate buffer is full.

Sizes of both the pre-fetch and write buffers can be specified during creation of the specialized iRODS Isilon resource. Preconfigured default sizes are used when corresponding parameters are omitted during creation of the resource.

LIMITATIONS OF THE DRIVER

Currently there is only one constraint[6] that is specific to the Isilon resource driver. Namely random writes. This is a natural consequence of the HDFS protocol limitations. HDFS does not allow random access for writing. Only streaming writes are supported[7].

Because of this limitation and specifics of current iRODS design, our solution demonstrates slightly lower write performance[8] of object upload[9] compared to NFS-based access to an Isilon. This is due to an artifact in how iRODS normally transfers data – using multi-stream I/O. It splits an object into many consecutive pieces and each piece is transferred to/from the storage resource by a dedicated thread. This manner (which is based on the POSIX I/O model) leverages 'random' (i.e. non-consecutive) writing which is not supported by our resource driver. Because of that we have to constrain object uploads to a single stream per object.

This limitation, while annoying, was deemed acceptable because:

1) Experience shows that data reads from an Isilon device under iRODS occurs much more frequently than writes. Optimizing to improve read speed is most valuable
2) Our preliminary performance evaluations show that the read (download) speed of our driver is significantly higher than that observed when using NFS access to the Isilon (see corresponding section below for details)
3) Preliminary data shows that in most cases the single-stream write performance of our driver is higher than single-stream performance of NFS-style writing. This observation suggests that aggregate write performance[10] will be higher in case of the specialized Isilon driver than in case of NFS-style writing

---

[5] More precisely, an individual buffer is allocated for each object handle opened through the driver. Several handles may correspond to a single object

[6] Other limitations exist, but they are not specific to our resource driver. For example, bundle iCommands are not supported by the driver. But any iRODS resource whose storage resource type is different from '**unixfilesystem**' will have this limitation. This characteristic is well-known, see [8] for the description

[7] In the case when random write access is a requirement, our solution may be used to form an 'archive' tier in a multi-tiered composite iRODS resources. E.g. a 'standard' iRODS '**compound**' resource may be created with the 'cache' tier represented by a '**unixfilesystem**' storage resource associated with an Isilon NFS mount, and an '**archive**' storage resource tier represented by the specialized Isilon resource driver

[8] 'Performance' here means 'time to complete an I/O operation measured at client side'

[9] Through '**iput**' iCommand

[10] In terms of the maximum possible throughput of multiple parallel uploads (measured at the client side)

4) Single stream I/O is not obliged in all practical cases to be slower than multiple stream I/O. Moreover, the single-stream approach has several advantages over multi-stream

The last point calls for clarification. Most client-side peripheral hardware resources may be kept busy by means of just single computing thread (i.e. one thread per one resource). For example, only one compute thread may keep a network adapter continually busy. The same is true for a disk drive if the thread is sending continuous I/O. This implies that if several hardware resources are required to complete an operation, the maximum performance may be achieved by using only one compute thread per resource. All these threads should keep their corresponding resources continually busy – but not overly busy! Thus, the number of threads required to achieve the best data transfer performance may be equal to the number of hardware resources involved in the transfer. For example, one thread reads data from a disc drive continuously and puts the I/O payloads in a queue. Then another thread takes the data from this queue and sends them over a network. And so on...

The important point here is that all I/O operations with hardware resources should be asynchronous. Otherwise the multi-stream model – involving multiple I/O streams per resource – may result in better performance than the described single-stream approach.

Unfortunately, data writing via HDFS is synchronous in the current HDFS implementations. But that should not be of great importance because the size of an HDFS IO payload may be chosen big enough to make the performance difference between synchronous and asynchronous protocols negligible.

The single-stream approach also results in sequential access to storage at both the client and Isilon sides. Sequential access is a preferred access mode for most storage systems. The single-stream approach allows better control over compute resources. During our investigation, we proposed to the Consortia to consider the implementation of a single-stream model – which better suits for modern object-based style of data access – in addition to the multi-stream one. We believe that many iRODS usage models – utilizing storage of various kinds – may benefit from this change.

LOAD BALANCING BETWEEN DIFFERENT ISILON CLUSTER NODES

Our data shows that the specialized Isilon resource driver results in almost perfect load distribution between different Isilon cluster nodes.

This better resources utilization occurs because Isilon's implementation of the HDFS NameNode service choses the appropriate Isilon node for each **individual** IO request.

Let us illustrate that with a graphic (Figure 1).

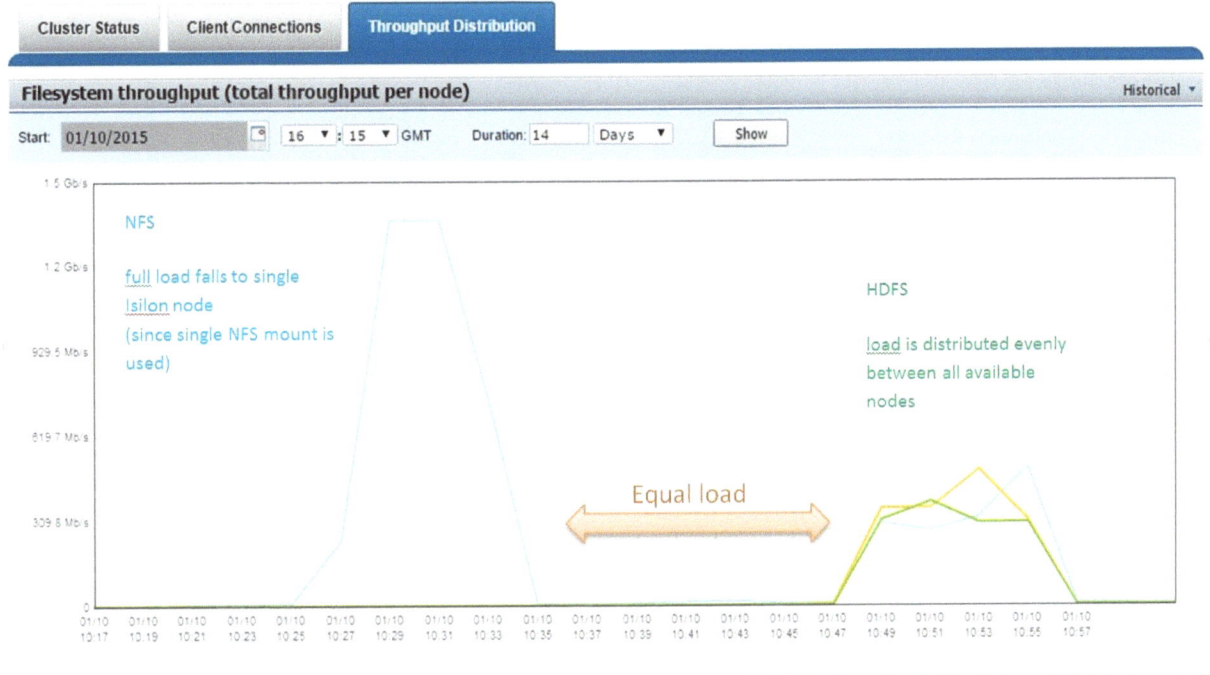

Figure 1. Load distribution among Isilon nodes in case of NFS-style and HDFS-style access

High peak on the left shows Isilon-side throughput for a particular set of iRODS operations[11] carried out using NFS-style approach. Three low peaks show Isilon's throughput for exactly the same experiment but carried out using the specialized Isilon resource driver in place of '**unixfilesystem**' driver. The resource driver and corresponding logical resource in iRODS are the only things that differ between the two experiments.

It may be seen that in the case of NFS the full I/O load falls to a single Isilon node, while in the case of the specialized resource driver, the load is distributed evenly among all available nodes (three in our case).

We want to stress that these results are preliminary. It may happen that intra-cluster limitations (e.g. limitations on traffic between cluster nodes) may not allow the I/O traffic to reach 100% of available cluster throughput under our new usage model. The data above should be considered as an illustration of the concept. Only stress-testing will provide complete proof of the concept. These works are planned but are not completed yet.

READ PERFORMANCE

We compared client-side performance of '**iget**' operation in the case of NFS-style access to Isilon and in the case of our solution. Our experiment was very simple:

1) We measured time required to download a single 4GB file from the Isilon
2) All that differed between the experiments was **just the plugin for File System access. All other parameters were exactly the same**
3) We measured how file download time depends on the size of intermediate buffer. In the case of NFS the size of the buffer may be altered through '*wsize*'/'*rsize*' parameters. In case of the specialized driver we used user-controlled sizing of the pre-fetch buffer

---

[11] 14 consecutive upload/download cycles performed for 2GB file

The simplest possible iRODS zone was configured to conduct the experiments (see Figure 2). The zone consisted of a single iRODS server which was an iCAT server. iRODS client commands – iCommands – were executed directly on this machine.

We used 10Gb Ethernet to connect the iRODS machine to the Isilon.

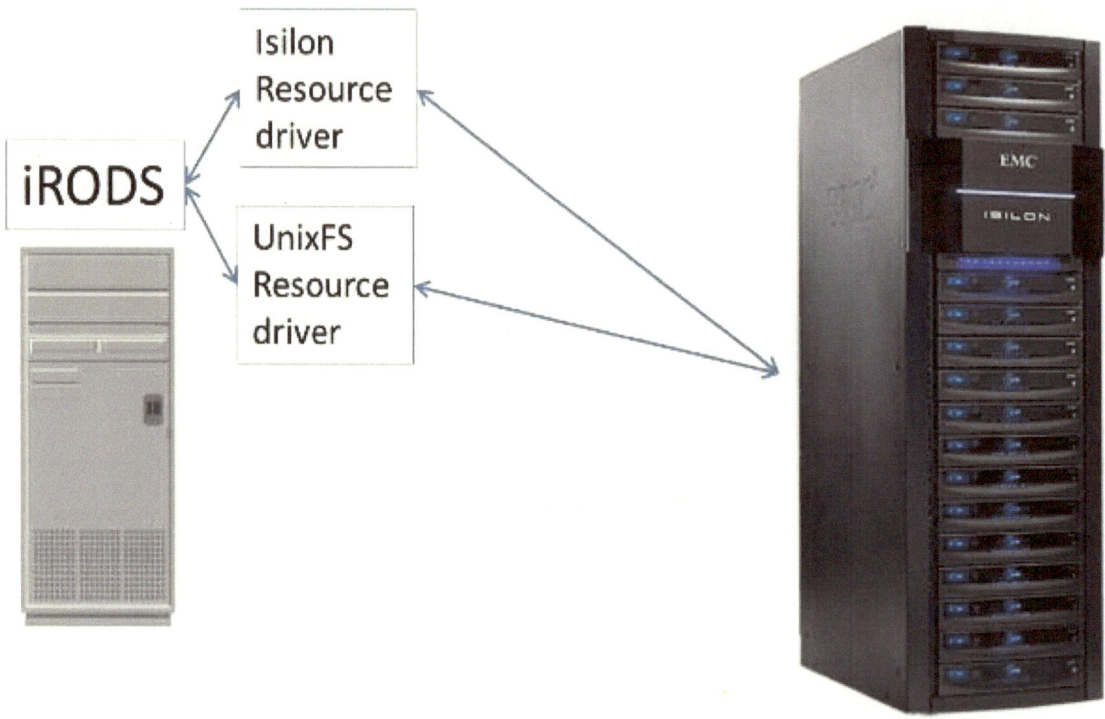

Figure 2. iRODS zone in our experimental setup

Results of the experiments are shown below (Figure 3). The X-axis shows size of the intermediate buffer in Megabytes. The Y axis shows time (in milliseconds) taken by the '**iget**' operation. For each buffer size and for each resource driver we made a series of 20 experiments. For the '**unixfilesystem**' driver – used to create iRODS resource over NFS – we plotted the minimum time observed in a series. In case of the specialized driver we plotted average time.

The reason we used 'average' for NFS is that we observed a huge inconsistency across a series of NFS experiments. As can be seen from the graph below, the minimum NFS time is about 7 seconds for all buffer sizes used[12]. But average NFS time is about 15 seconds for all buffer sizes used. Currently we cannot explain this tremendous inconsistency across NFS experiments. That's why we use minimum observed time for NFS.

In case of the specialized plugin, the minimum observed time doesn't differ from average by more than 10%. This is why we are comfortable using the 'average' in this case.

---

[12] Here we ignore high peaks resulted from random fluctuations

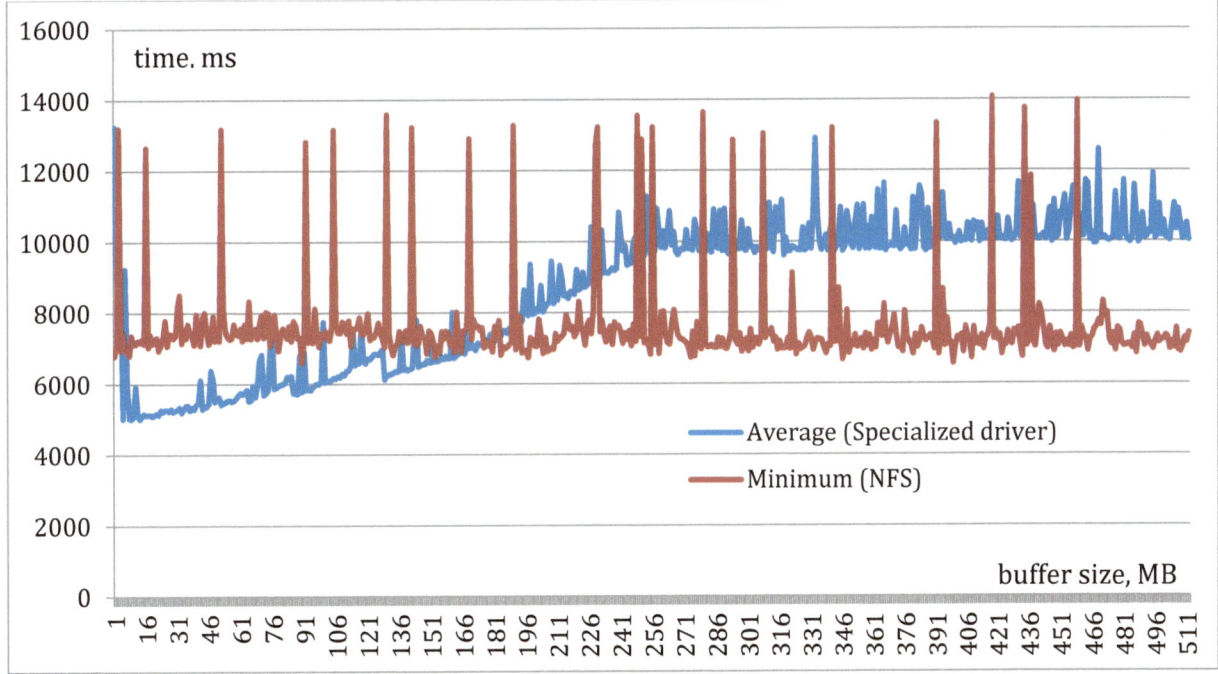

Figure 3. Performance of NFS-style access to Isilon compared to performance of the specialized driver

It can be seen from the graph that NFS performance doesn't depend significantly on the size of the intermediate buffer. Actually it does, but this dependence can be observed at KB scale only (for buffers < 1MB). The graph above was built at MB scale. NFS performance for KB-sized buffers is lower than that for MB-sized buffers.

As follows from the graph, the best NFS performance (~7 seconds) is about 30% lower than best performance of the specialized driver (~5 seconds for buffer sizes close to 10MB).

It is interesting to note that in our experiments it turned out that the 4GB file was downloaded into memory. A file wasn't flushed to a local disk when '**iget**' iCommand completed. Our local disk has write performance of about 100MB/s, while maximum performance observed in the experiments is about 800MB/s.

While the observed performance difference is promising, the results should be treated with care. The experiments are very simple. Stress-testing is required to understand real performance potential of the specialized Isilon resource driver. Multiple iRODS clients operating on multiple iRODS servers simultaneously should be used to make a good case for performance evaluations. We are working on that now.

We are working to understand NFS performance limitations. Performance bottlenecks of the specialized driver should also be studied. Currently the best observed performance of the specialized driver in terms of client-side bandwidth is still 35% less than the maximum performance of 10Gb network link. Also it's 67% less than the performance of memory[13].

CONCLUSIONS AND FUTURE WORK

Our initial evaluations of the specialized iRODS Isilon resource driver shows that nearly perfect resource balancing at the Isilon side can be achieved. Also client-side read performance in a very simple usage case can be 30% better than in case of conventional NFS-style access to Isilon.

---

[13] We measured this performance in the following way:

```
sudo mount -t tmpfs -o size=9000m tmpfs /mnt/ramdisk
time dd if=/dev/zero of=/mnt/ramdisk/memtest bs=4M count=1024
```

Nevertheless all the evaluations presented are preliminary. They were developed for very simple use cases and should be treated as 'concept illustration' and not as 'concept proof'. Stress-testing is necessary to prove or disprove the concept. This work is underway.

Besides performance evaluations of the new resource driver, we are considering further functional and performance improvements:

1) Intelligent read pre-fetch and write bufferization in the background
2) Support of Isilon's advanced features
3) Better customer experience via easier setup

During our work on the Isilon-specific resource driver we investigated the data-transfer aspects of iRODS. This resulted in a number of issues related to stability, usability, and performance being reported.

ACKNOWLEDGEMENTS

We are very grateful to Patrick Combes, and Sasha Paegle. These people along with Steve Worth initiated the work on the specialized support of Isilon in iRODS. Their support and guidance are priceless for us.

REFERENCES

[1] EMC Isilon For Life Sciences, http://www.emc.com/collateral/solution-overview/h11212-so-isilon-life-sciences.pdf

[2] Charles Schmitt, RENCI, "iRODS for Big Data Management in Research Driven Organizations". https://abc.med.cornell.edu/assets/irods_workshop_schmidt.pdf

[3] Bedard, Dan: Managing Next Generation Sequencing Data with iRODS, http://irods.org/wp-content/uploads/2014/09/Managing-NGS-Data-using-iRODS.pdf, Visited last on 26.05.2015

[4] IDC MarketScape Excerpt: Worldwide Object-Based Storage 2013 Vendor Assessment. http://russia.emc.com/collateral/analyst-reports/idc-marketscap-ww-object-based-storage-vendor-assessment.pdf

[5] iRODS GitHub, https://github.com/irods, Visited last on 26.05.2015

[6] EMC press release from January 31, 2012, http://www.emc.com/about/news/press/2012/20120131-01.htm, Visited last on 26.05.2015

[7] Hadoofus, https://github.com/cemeyer/hadoofus, Visited last on 26.05.2015

[8] iRODS GitHub issue dedicated to virtualization of bundle operations, https://github.com/irods/irods/issues/2183, Visited last on 26.05.2015